Introduction to
STATISTICAL ANALYSIS
(With some SAS procedures)

Malik B. Malik, Ph.D.

Department of Mathematics

School of Education, Social Sciences, and the Arts
University of Maryland Eastern Shore
Princess Anne, MD 21853

First Edition

ISBN: 979-8-89228-918-4   (Paperback)
ISBN: 979-8-89228-919-1   (eBook)

Printed in the United States of America

Author's Dedication

With humility and utmost respect, I dedicate this piece of work to my Spiritual Leader and my Mentor in my journey to eternity; **Elustaz: Mahmoud Mohamed Taha.**

Preface

This edition of the Handbook emanated from a collection of lecture notes delivered to graduate students at the University of Maryland Eastern Shore throughout many years to help them in their research methodology. The handbook's focus is to take students with very little basic statistics into more advanced statistical analysis encouraging statistical thinking among students who are only interested in understanding the subject of statistical analysis.

The handbook does not cover probability theory in any depth, although we know that statistical analysis's foundations are rooted in probability and probability distributions. For that reason, we had an introduction to elemental distributions that are used extensively in statistical analysis for readers to understand how these distributions fit into the statistical inference. Namely, we exposed the reader to Fisher F-distribution, Student t-distribution, and Chi-square distributions and how they relate to the Normal Distributions and to each other.

There are some practice problems for students to check on their understanding of the material; we are hoping that in the near future, we may supplement this handbook with a website that exposes students to more applied SAS-based projects and practical programs. At this stage we refer students to a very useful open source with the following URL:

http://www.ats.ucla.edu/stat/sas/

There are a lot of SAS, as well as other software programs to help users use any of the available programs and modify them to serve their analytical and inferential purposes. We find this website really useful and with a lot of self-explanatory pieces of useful code and programs. The site also helps readers learn and get started with many of the different applications of statistical software, including SAS, R, SPSS, and STATA.

Statistics is a very applied subject and, for some, just research techniques and methodology, but there is no denying that it is also abstract. Our objective here is to present it in a way that anyone exposed to post-calculus algebra can understand without spending many background semesters preparing.

We presented statistical concepts that are meant to review basic statistics in the first chapters. It was also our belief that statistics cannot be complete without some exposure to computational software, and that is why we added some SAS programs that serve the contents of this book directly, yet we are aware that software exposure needs learning a foundation of statistical packages like SAS, R, STATA, SPSS, etc., Which is not provided in this book.

We believe that it is never easy for the handbook to fully and comprehensively serve all the objectives set out at the beginning within its first print or edition. We will immediately be working on editing the handbook and on providing a website to complement this handbook in its next edition, considering this first print as a work in progress. Feedback from students, instructors, colleagues, and experts in the wider professional arena will be more than welcome to critique and thus improvement of this work.

Malik Beshir Malik, Ph. D.

December 1, *2022*

Department of Mathematics and Computer Science

University of Maryland Eastern Shore

Princess Anne, MD 21853

Chapter One
Introduction

1.1. Historical Note

The history of simple forms of statistics emerged at the beginnings of civilization when symbols or pictorial representations were used to document population sizes, numbers of livestock, amounts of monies, taxes, sizes of armies, etc. The Babylonians used clay tablets for agricultural records, especially during Hammurabi's reign (18th Century BC). We can consider the biblical books of Numbers and Chronicles as statistical reporting censuses of the Israelites and their tribes' material wealth. Similar records were reported in Chinese history. The Greeks in the 7Th Century BC used state data for defence, war, and taxation purposes. Counting principles may be the only mathematics needed for ancient "statistics" [1]. The Roman Empire gathered extensive data on population and wealth. The Carolingian kings Pepin in 758 AD and Charlemagne in 762 AD ordered surveys of holdings. These are just historical examples of ancient "statistics." Some scholars state that the 1662 publication of Natural and Political Observations by John Graunt is the origin of statistics. It must be mentioned that most of the statistical concepts originated in non-mathematical settings, like the idea of a range, measures of centrality, and dispersion, to name a few, but were late firmed up and further developed using the rigor of mathematics like linear algebra, measure theory using Riemann–Stieltjes integrals.

The mathematics of statistics may have started with Thomas Bayes in the mid-18th century with interests in games of chance ("win versus lose") and soon found applications in all facets of life. Some other historian statisticians attribute the mathematical developments of statistics to the 17th Century development based on probability theory to Blaise Pascal and Pierre de Fermat. Carl F. Gauss was cited as the first to describe the least square method in 1794. Statistical names emerged in the first half of the 20th century, like K.A. Fisher, J. Neyman, A. Wald, and E.S. Pearson, followed by R. Schlaifer, J. Savage, and other statistical theoreticians. Names like Spearman, Gosset, R. A. Fisher, De Morgan, Boole, Laplace, and others must be cited when looking into modern statistics.

1.2. Statistics and Computers

In the second half of the 20th century, minicomputers and mainframes revolutionized statistical analysis. With the 1980s invention of personal computers, statistical analysis became more available and accessible to scientists, researchers, and students. That was reflected in the development of much platform-transparent statistical software, which is getting cheaper and accessible for "small data."

Statistics as science grew exponentially in content, versatility, and use with computers. Such growth of statistics lent itself to almost every facet of human activity. Before the computer revolution described above, statistics could not collect data and make an introductory presentation in charts and tables. However, nowadays, statistics is widely encompassing the science of inductive inferences. Entire activities of making decisions in the face of uncertainty and risks are becoming easier to entertain. Thanks to the dependent relation of probability and statistics, an engineer quality control unit accept, rejects, or corrects a production process, a dietician can experiment with food additives, an economist forecasts trends, a public opinion poll predicts an election, and an actuary scientist decides life insurance premiums, and so on so forth. Previous "Small data" statistical analysis is also moving into data mining and "big data" analytics, which reflects another leap in taking advantage of opportunities of analysing the vast voluminous data sets made available by the Internet beyond the imagination of early statistical theoreticians.

1.3. Statistical Thinking: Preparations and Challenges

Statistical theory and inductive inference have played an enormous part in developing science and knowledge, not to downplay the complementary advancement of controlled laboratory experiments. Based on sampling and samples, statistical theoreticians made it possible through inductive inference to generalize from particular incidences to the whole using probabilistic models, thus adding to knowledge and discovery in science (human and exact) every day. Knowledge and scientific discovery, besides the controlled laboratories as a method, would also need the statistical way of thinking to the readers of this book. However, since the probability is a semester-long, we would not expect the readers to be delighted with the contents of this book to directly dive into research without some help from further reading or thorough coaching by a mentor and sometimes a prerequisite for laboratory investigation. This section will not cite the cases where a scientist needs statistics.

However, we must emphasize the need for statistical investigations and analysis in every aspect of human life.

One important point we need to make clear is that scientists in many fields need not only to understand statistics and statistical thinking but also to be able to use statistical (computational) software.

The two sides of this "coin" of computer software and statistical analysis are no longer separable. That is why we are not separating statistics from computational software in this book. We are challenged, and still, in making this book serve that goal. The double challenge is that learning statistics is by itself challenging, and we were sensitive not to complicate the learning of statistics with the learning of software. These should be taught separately as statistics lectures, followed by computational laboratory sessions.

The book's layout has been strategically geared to start with basic statistical inference (assuming that a course of elementary descriptive part is already studied), which lays the foundation for statistical thinking. Along with that, an introduction to computer software is recommended. This portion (or module) can take the first half of the semester. The second half of the semester witnesses the heavier and relatively more advanced module that deals with concepts of analysis and variance, multiple regression, and factorial analysis as critical statistical concepts that every researcher must know. The book is meant to introduce the statistical way of thinking to the reader, but since in just a semester-long course, we would not expect readers to be completely content with the contents of this book to dive into research. They must have some help by further reading or special coaching through a mentor (usually a faculty or consultant in research institutions like universities).

Although the statistical theory and principles are always the same, we must make everyone aware that statistics can be assimilated a bit differently in the different and diverse sciences to the extent that the statistical flavour implemented in biology is not the same as in, say, economics or physics. In this respect, we always advise students to read their discipline-specific journals or publications to see how they used statistics in their field and to use the language of statistics spoken amongst their own peers, mentors, and pioneers.

Another remark is that this book assumes that the reader has been exposed to probability because probability is critical to statistical inference. However, the probability is not explicitly addressed in the book. But it suffices to say that the concepts of "sampling," "distributions," and "random variable," which are mentioned here a lot, are inherently probabilistic concepts.

For this reason, we have developed an introduction section to define and describe the quoted concepts above and tie them to probability.

1.4. Sampling and Random Variables

We start by making a clear distinction between a population and a sample. The population is generally defined as all the elements or observations under investigation (e.g., all students on campus at a given time) whose particulars like age, income, grades, etc., are being studied} and the researcher would like to make conclusions about their behaviour. In any study, a researcher may be interested in generalizing the whole population. Still, for reasons of expense or impossibility, the researcher may not be able to cover each and every element of the population. The option is to look into a part of that population (referred to as the sample) and then, through principles of statistical theory and methods, make inferences about the whole population using that subset (sample) of the population. Randomization and replication are very important concepts in applied statistics. Here below, we give an idea about these concepts.

Statistical Randomization can be simply defined as the random process of assigning treatments to experimental units. It is the practice of using chance methods to assign subjects to treatments such as flipping the coin. It allows for using probability theory to express the likelihood of chance. Probability is the ratio of times an event occurs to the total number of trials. Randomization ensures that each patient has an equal chance of receiving any treatment under a study. Randomization ensures that selection bias is eliminated. A randomized experiment is an important tool for testing the efficacy of a treatment.

There are three popular methods of Randomization; they are simple, block, and stratified. Simple Randomization is a simple and easy way to implement Randomization. It is a simple way of randomly assigning subjects to different treatments. A common method of this Randomization is flipping a coin; others include a shuffled deck of cards and throwing dice. For example, Heads may equal active treatments and tails for placebo treatment. The assignment of treatment is completely unpredictable.

Block randomization comes into place because simple Randomization does not guarantee balance in numbers during the trial, especially when patient characteristics change with time, which means that early detection of imbalances cannot be corrected. Block randomization is often used to fix this issue.

Stratified Randomization can balance subjects on baseline covariates, which tend to produce comparable groups with regard to certain characteristics (such as gender, age, race, and disease severity), thus producing valid statistical tests. Subjects should, however, have baseline tests taken before Randomization.

In conclusion, the benefits of Randomization include eliminating selection bias, and it balances arms with respect to unknown and known variables and forming the basis for a statistical test of the equality of treatments.

Replication means the repetition of an experiment and observation in the same or similar conditions. It could also be described as the repetition of the basic experiment, where the experiment is performed more than once. Repeating each treatment to more than one unit, a more reliable estimate of the effect of each treatment can be attained. Replicated observation of the experimental units can be used to determine the importance of the treatment difference.

In statistics, replication is the repetition of an experiment or observation in the same or similar condition. Replication is important because it adds information about the reliability of the conclusions or estimates to be drawn from the data given. It can also involve applying the theory to new situations in an attempt to determine the generalizability to different age groups, locations, races, or cultures.

Replication means making more than one measurement at the same combination of factor levels. In order to count as proper replication, all potential sources of random variation must be allowed to operate independently on each occasion. Replication can also be a convenient way of increasing the size of the experiment and hence increasing precision.

It is important to design an experiment that can be replicated because; some results may have been skewed or wrong. Doing multiple trials helps provide assurance that the results are correct, and it also allows you to get an average in some areas observed results are likely to be affected by random chance.

There are two types technical and biological. Technical replication is repeated measurement of the same sample, which is important in controlling for errors in measurement.in biological replication, measurements are taken from several independent biological subjects rather than from simple individuals.

Replication is important for a number of reasons, including assurance that results are valid and reliable, determination of generalizability or the role of extraneous variables, applications of results to real-world situations, and inspiration of new research combing previous findings from related studies.

1.5. A Convention

It is a convention in statistics to denote parameters using Greek letters and sample estimators or values using English letters. For example, a population mean is denoted by μ while the sample means can be denoted by.

The word "statistics" really developed into today's concept that is basically probabilistic and limited to sampling, yet the ancient uses (if any) are basically census and population based.

Chapter Two
The Normal Distribution and the Chi-Square, t, and F-distributions

In this chapter, we will learn about the Normal Distribution (sometimes known as the Gaussian Distribution) and show its relation to the Chi-square, the t, and the Fisher F distributions and how the latter sampling distributions emanate from the Normal Distribution. We will show how these sampling distributions are constructed based on a random sample from a Normally distributed population as a ground-breaking for how many of the statistical methods and techniques are directly based on all the above-mentioned distributions.

In this book, and without any loss of generality, we will assume that a random sample of size n is drawn from a population that is Normally distributed with mean μ and variance σ^2 (sometimes denoting that by $N(\mu, \sigma^2)$. We also denote two independent samples from two populations $N(\mu_1, \sigma_1^2)$ and $N(\mu_2, \sigma_2^2)$ by and respectively.

2.1. The Normal Distribution

Many statistical techniques in statistical inference are based on the Normal Distribution, which is a bell-shaped symmetrical distribution with mean μ and variance σ^2 for a rand variable X such $-\infty < X < \infty$ and σ is finite.

In real life, we may get our sample and represent our data in histograms which might reflect a non-symmetrical shape. Sometimes the histogram shapes might look as in the example below:

There are many cases where the data tends to be around a central value with no bias left or right, and it gets close to a "Normal Distribution" like this:

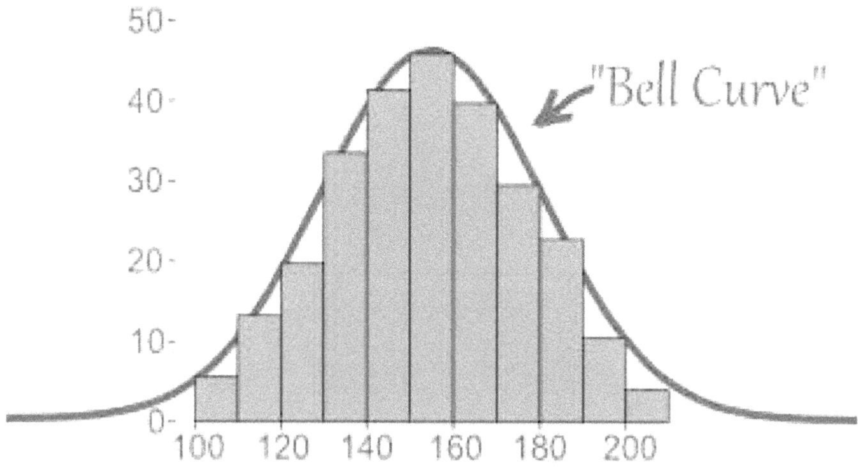

Fig 1. A Normal Distribution

The "Bell Curve" is a Normal Distribution.

And the yellow histogram shows some data that

follows it closely but not perfectly (which is usual).

In such cases, we can assume that our data is Normally Distributed and use the sample mean to estimate its mean μ and the sample standard deviation s to represent its σ.

Many things closely follow a Normal Distribution:

- heights of people
- size of things produced by machines
- errors in measurements
- blood pressure
- marks on a test

We say the data is "normally distributed":

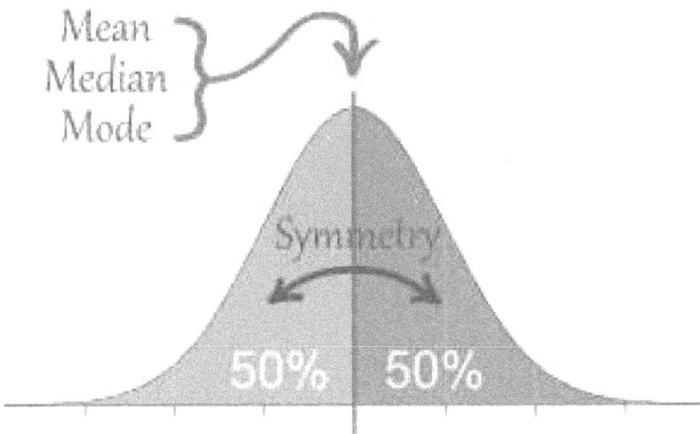

Fig 2: The Normal Distribution curve showing symmetry.

As is clear from the Fig 2, the Normal Distribution has the following properties:

- mean = median = mode
- symmetry about the center
- 50% of values less than the mean and 50% greater than the mean

Standardization (Transform of a general N(μ , σ2) into a N(0 , 1)):

A general N(μ , σ²) random variable X can be transformed into a Standard N(0, 1)

Z-variable by using the standardization formula:

$$Z = \frac{x - \mu}{\sigma}$$

(A)

which is used whenever we are standardizing (transforming) one single X value

(B) $$Z = \frac{\bar{x} - \mu}{\sigma / \sqrt{n}}$$ (same as saying $$Z = \frac{\sqrt{n}.(\bar{x} - \mu)}{\sigma}$$)

which is used whenever we are standardizing (transforming) he sample mean.

The justification for (A) and (B) is that if X is N (μ, σ^2), then is N (μ, $\dfrac{\sigma^2}{n}$). That is to say, when the

Var(X) = σ^2 then the Var ($\overline{X}$) = $\dfrac{\sigma^2}{n}$

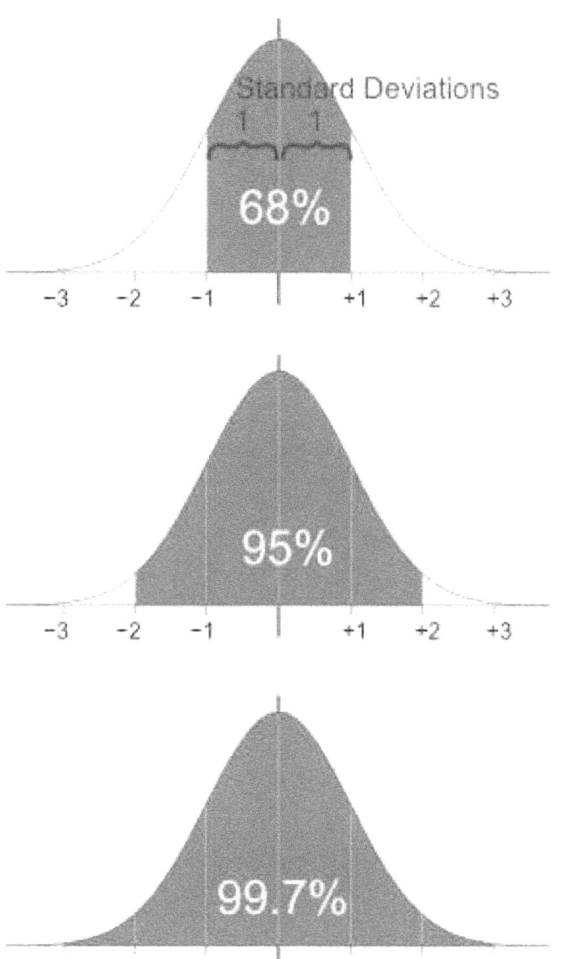

68% of values are within
One standard deviation of the
mean

95% of values are within
Two standard deviations of
the mean

99.7% of values are within
Three standard deviations of
the mean

Fig 3:   Standard Normal Probability Distribution

We note that as a consequence, the mean of Z is 0, and the Var(Z) =1. In other words, Z is distributed as the Standard Normal probability distribution, N (0, 1) and is tabulated in almost every book of statistics and is useful in finding any probability for any Normal Distribution.

When we <u>calculate the standard deviation</u>, we find that (generally):

2.2. The t-distribution

The f-distribution is very similar in shape to the normal distribution but works better for small samples when σ is not known. In large samples, the f-distribution converges to the normal distribution.

Properties of the t-distribution

In the previous section, we explained how we could transform a normal random variable with an arbitrary mean and an arbitrary variance into a <u>standard normal variable</u> (with mean 0 and variance 1, i.e. N (0, 1)). That was on condition that we knew the values of the population parameters. Often it is not possible to know the population variance, and we have to rely on the sample value of variance. The transformation formula would then have a distribution that is different from the normal in small samples. It would instead be f-distributed.

We use the transformation formula to form the test function.

1. The central t-distribution is symmetric around its mean.

2. The mean equals zero, just as for the standard normal distribution.

3. The variance equals k/(k-2), with **k** being the degree of freedom.

$$t = \frac{\overline{X} - \mu_X}{\frac{S}{\sqrt{n}}} \sim t_{(n-1)}$$

Observe that the expression for the standard deviation contains an S. S represents the sample standard deviation. Since it is based on a sample, it is a random variable, just like the mean. The test function, therefore, contains two random variables. That implies more variation and, therefore, a distribution that deviates from the standard normal. It is possible to show that the distribution of this test function follows the t-distribution with **n-1** degrees of freedom, where **n** is the sample size.

2.3. The Chi-square distribution

Until now we have talked about the population mean and performed tests related to the mean. Often it is interesting to make inferences about the population variance as well. For that purpose, we are going to work with another distribution, the Chi-square distribution.

Statistical theory shows that the square of a standard normal variable is distributed according to the Chi-square distribution, and it is denoted by $\chi^2_{(1)}$ (read as "Chi-square with one degree of freedom" and has one degree of freedom since we squared one standard normal. It turns out that the sum of squared independent standard normal variables also is Chi-squared distributed with degrees of freedom equal to the sum of the individual Chi-squared random variables. We have:

$$Z_1^2 + Z_2^2 + \ldots + Z_k^2 \sim \chi^2_{(k)}$$

And if W and R are Chi-squared random variables with k and r degrees of freedom, respectively, then the sum.

$$W + R \quad \sim \quad \chi^2_{(k+r)}$$

Properties of the Chi-squared distribution

1. The Chi-square distribution takes only positive values (since it is a squared or a sum of squared variables)

2. It is skewed to the right in small samples and converges to the normal distribution as the degrees of freedom go to infinity

3. The mean value equals k, and the variance equals 2k, where k is the degrees of freedom

In order to perform a test related to the variance of a population using the sample variance we need a test function with a known distribution that incorporates those components. In this case, we may rely on the statistical theory that shows that the following function would work:

$$\frac{(n-1)S^2}{\sigma^2} \sim \chi^2_{(n-1)}$$

Where S^2 represents the sample variance, σ^2 the population variance, and n-1 the degrees of freedom used to calculate the sample variance. Later on, we will see how this function could be used to perform a test related to the population variance.

2.4. The F-distribution

The final distribution to be discussed in this chapter is the F-distribution. In shape, it is very similar to the Chi-square distribution but is a construction based on a ratio of two independent Chi-squared distributed random variables descaled by their respective degrees of freedom. An *F-*distributed random variable, therefore, has two sets of degrees of freedom since each variable in this ratio has its own degrees of freedom (one set for numerator degrees. That is, if W and R are Chi-squared random variables with k and r degrees of freedom, respectively, then:

$$\frac{W/k}{R/r} \sim F_{(k,r)}$$

Note that the reciprocal of this ratio will correspond to a complement of a distribution.

Properties of the F-distribution

1. The F-distribution is skewed to the right and takes only positive values (since it is a ratio of positive functions).
2. The F-distribution converges to the normal distribution when the degrees of freedom become large
3. The square of a t-distributed random variable with k degrees of freedom becomes an F-distributed random variable with (1, k) degrees of freedom

The F-distribution can be used to test the equality of population variances. It is especially interesting when we would like to know if the variances from two different populations differ from each other. The statistical theory says that the ratio of two sample variances forms an F-distributed random variable with n_1 -1 and n_2 -1 degrees of freedom when the population variances are assumed equal (i.e., $\sigma_1^2 = \sigma_2^2$, that is:

$$\frac{S_1^2}{S_2^2} \sim F_{(n_1-1)(n_2-1)}$$

The reason for this is two-fold:

First:
$$S_2^2 = \frac{\sum_{i=1}^{n_2}(x_{2i} - \bar{x}_2)^2}{n_2 - 1}$$

Second, we know that $\dfrac{\sum_{i=1}^{n_1}(x_{1i} - \bar{x}_1)^2}{\sigma_1^2}$ and are $\chi_{(n_1-1)}^2$ **and respectively.**

Theorem (without a proof):

F_{α,n_1-1,n_2-1} is the reciprocal of $F_{1-\alpha,n_2-1,n_1-1}$

where α and $1-\alpha$ both represent the upper-tail probability level of the F-distribution (noting that the degrees of freedom n_1-1 and n_2-1 have been interchanged to reflect the reciprocal characteristics.

2.5. The Z, t, Chi-square, and F distributions in statistical Inference

The random variables Z, t, χ^2, and F are essential for developing some of the statistical inference procedures and models. Later on, we will see how they are used in hypothesis testing, confidence interval estimation, analysis of variance, multiple regression, and design of experiments. All of these will require the computation of probability. All the above-mentioned probability distributions are well-tabulated in statistics textbooks.

Chapter Three

Review of Hypothesis Testing for Means and Variances

A <u>statistical hypothesis</u> is a statement about a characteristic of one or more populations. The hypothesis may be <u>true</u> or <u>false</u>. From sample data, you want to <u>reject</u> or <u>fail to reject</u> the hypothesis. This is accomplished by testing to see if the <u>sample statistic</u> differs <u>or</u> does not differ significantly from the hypothesized <u>parameter</u>.

(Note: If the population mean, μ, and population standard deviation, σ, are unknown, they must be estimated from sample data (s).)

There are four steps in hypothesis testing:

(1) State H_o and H_A where H_o is called the <u>null hypothesis</u> and H_A is called the <u>alternate hypothesis.</u>

(2) Determine the decision rule -i.e., decide on an α level.

(3) Take a sample and compute the test statistics.

(4) Draw the conclusion, i.e., reject or fail to reject H_o

(Note: The ☐ level determines the critical values and the rejection regions.)

Hypothesis tests can either be two-tail or one-tail. In a two-tail test, we reject H_o if the test statistic is either too large or too small. A one-tail test can either be an upper-tail or a lower-tail test. In an upper tail test, we reject H_o when the test statistic is too large. In a lower tail test, we reject H_o when the test statistic is too small.

In hypothesis testing, we can make mistakes. Depending on the value of the sample statistic, we can reject a true H_o (Type I or α error), or we can fail to reject a false H_o (Type II or α error).

In tabular form:

	H_0 **True**	H_0 **False**
Fail to Reject H_0	Correct Decision	Type I error
Reject H_0	Type II error	Correct Decision

We must also be aware of whether our <u>sample size</u> is <u>large</u> ($n \geq 30$) or <u>small</u> ($n < 30$) and whether the <u>population standard deviation</u> is <u>known</u> ($\square$) or <u>unknown</u> (s) (estimated from a sample). These facts determine whether we use a <u>z</u> or a <u>t</u> distribution.

1-1

In tabular form:

	σ known	σ unknown (use s)
n ≥ 30	z	z
n < 30	z	t

Let's do three examples to illustrate hypothesis testing.

Ex 1: Forty-nine American soldiers observed at random yield a mean weight of 160 lbs., with a standard deviation of 11 lbs. Are these observations consistent with the assumption that the mean weight of all American soldiers is 170 lbs. at $\alpha = .01$?

$n = 49 \quad (n \geq 30) \quad \rightarrow \quad z$

$s = 11$ (σ unknown) $\nearrow$

$\bar{x} = 160$

$\alpha = .01$

Now for the four hypothesis testing steps:

(1) H_o: $\mu = 170$

(2) H_A: $\mu \neq 170$ $\rightarrow$ two-tail test

(3) $Z_{\alpha/2} = Z_{.005} = 2.58$

(4) (Remember, you look up $.5 - \alpha/2 = .495$ in the z table)

Since this is a two-tailed test, our decision rule is: Reject H_o if $z^* < -2.58$ or if $z^* > 2.58$

Let's draw a picture of the decision rule:

(3) Test statistic is

α

$$z^* = \frac{160 - 170}{11/\sqrt{49}} = -6.36$$

(4) Compare -6.36 to the picture above. Where does it fall? Left tail rejection region. The conclusion is to reject the H_o at $\alpha = .01$. Based on our sample data, the soldiers' average weight is not 170 pounds. In addition, since we rejected the left tail, we can say that the soldiers probably weigh less than 170 pounds. (If we had rejected in the right tail, we would say that the soldiers probably weigh more than 170 pounds.)

Ex 2: We feel that the IQ's of the freshman at SMS are above the national average of 100. According to past knowledge, IQ is normally distributed with a standard deviation of 10. A random sample of

25 freshmen is selected, and a mean IQ of 104 is observed. Do the IQs of freshmen at SMS run above the national average at $\alpha = .01$?

$n = 25$ (n<30) $\rightarrow \rightarrow \rightarrow \rightarrow$ z

$\sigma = 10$ (σ known because of historical information) $\nearrow$

$\alpha = .01$

$\bar{x} = 104$

(1) H_o: $\mu \leq 100$

(2) H_A: $\mu > 100$ $\nearrow$ one-tail test (upper tail)

Note: The <u>equality sign</u> always goes in H_o. We are asked whether the freshmen are <u>above</u> the average of 100, i.e., $\mu > 100$. Since this is a strict inequality (no equal sign), it appears in H_A. The complement $\mu \leq 100$ appears in H_o.

Note: In a one-tail test, if H_A is in the order "Greek letter, inequality symbol, the number," and the symbol is >, it is an upper-tail test. If the symbol is <, it is a lower tail test.

(2) $z_\square = z_{.01} = 2.33$

(Remember, this is <u>one-tail,</u> so we <u>don't</u> divide α by 2. Look up .5- α =.49 in the z table.) Reject H_o if $z^* > 2.33$. This is an upper-tail test, so we will only have a rejection region in the upper tail.

Let's draw the picture:

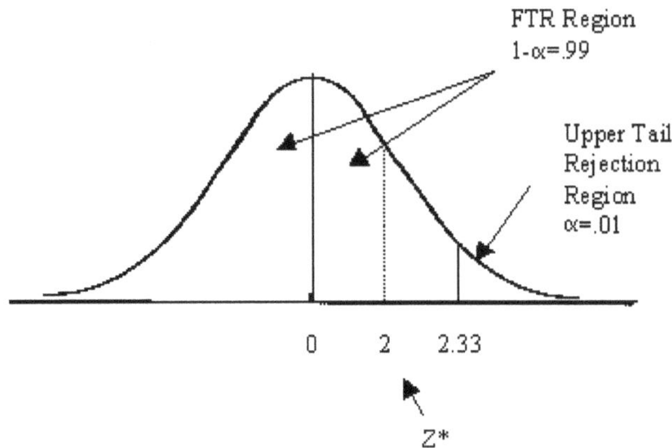

(3)

$$z^* = \frac{\overline{X} - \mu_0}{\sigma/\sqrt{n}} = \frac{104 - 100}{10/\sqrt{25}} = 2$$

Draw z^* in the picture (above). It falls in the fail-to-reject (FTR) region.

(4) FTR H$_o$. From our sample data, we <u>cannot</u> conclude that the freshman IQs run above the national average at α = .01.

Note: Our 1st example was two-tail because we wanted to know if the soldiers weighed 170 lbs. or not, i.e., they could have been heavier or lighter than 170 lbs. Our 2nd example was one-tail because we wanted to know if the freshman were <u>above</u> average.

<u>Example 3</u>:   The City of Springfield uses tens of thousands of light bulbs every year. The brand that has been used up to now has a mean life of 1000 hours with a standard deviation of 90 hours. A new brand is offered to the city at a much lower price. The city decides that the new brand should be purchased now unless it has a mean life of fewer than 1000 hours at □=.05. Subsequently, 25 bulbs of the new brand are tested, yielding an average life of 990 hours with a standard deviation of 80 hours. What should the city decide?

n=25 (n<30) $\rightarrow$ $\rightarrow$ $\rightarrow$ $\rightarrow$ t with 24 degrees of freedom (n-1)

s=80 (□ is unknown) $\nearrow$

α =.05

$\bar{X} = 990$

(1) H_o: $\mu \geq 1000$

(2)　　　　H_A: $\mu < 1000$　　$\rightarrow$ one tail test (lower tail)

H_o indicates that the bulbs are at least as good as the old brand; i.e., they have a life of at least 1000 hours. H_A indicates that the bulbs are inferior, i.e., the life is less than 1000 hours. Remember: Equality always goes in H_o!

(3) $t_{\alpha,\,n-1} = t_{.05,\,24} = -1.711$

(4) (Remember this is a one-tail test, so we don't divide α by 2, look up $t_{.05}$ with 24 degrees of freedom in the t table.)

Since this is a lower tail test, reject H_o if $t^* < -1.711$.

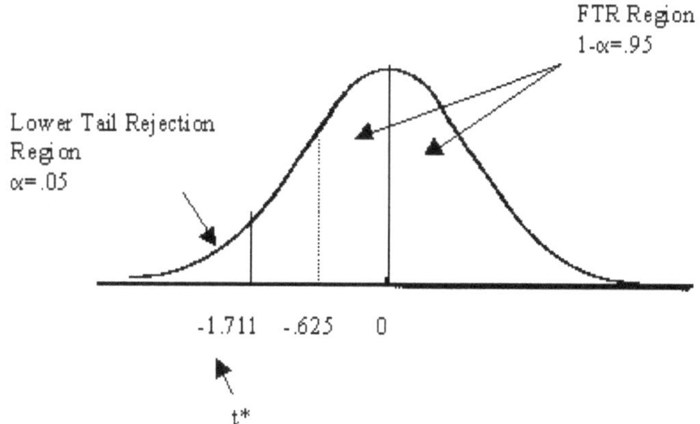

(3)

$$t^* = \frac{\bar{X} - \mu_0}{s/\sqrt{n}} = \frac{990 - 1000}{80/\sqrt{25}} = -.625$$

In the picture, $t^* = -.625$ falls in the FTR region.

(5) FTR H_o at $\alpha = .05$. Our sample data indicate that the City should buy the new, cheaper brand of bulbs.

Hypothesis Testing vs. Confidence Intervals

Confidence intervals yield the same information as hypothesis testing. For example, the upper and lower limits for a confidence interval for the population mean μ re-derived using the formula. If μ_o lies <u>within</u> the confidence limits, we <u>fail to reject</u> H_o. Otherwise, reject H_o.

Ex: Let's use our "49 American Soldiers" example.

$n = 49 \ (n \geq 30) \qquad \rightarrow \quad z$

$s = 11 \ (\sigma \text{ unknown})\square \nearrow$

$\alpha = .01$

$\bar{X} = 160$

Generate the confidence interval.

$$\bar{x} \pm z_{\alpha/2} \, s/\sqrt{n}$$

$$160 \pm z_{.005}(11/\sqrt{49})$$

$$160 \pm 2.58(1.57)$$

$$160 \pm 4.05$$

$(155.95, 164.05)$ is the 99% Confidence Interval

i.e. $P(155.95 \leq \mu \leq 164.05) = .99$

Now recall H_o and H_A

$H_o: \mu = 170$

$H_A: \mu \neq 170$

Is $\mu_o = 170$ contained in the confidence interval? No. Therefore, we must reject H_o. This is the same conclusion that we reached earlier in hypothesis testing.

With CI's (confidence intervals), you actually get more information than with hypothesis testing. We can look at our CI and tell that not only do the soldiers <u>not</u> weigh 170 lbs, but we are also 99% confident that they weigh, on average, between 155.95 and 164.05 lbs.

Note: If you have a one-tail hypothesis test, you generate a two-sided CI by doubling the α value.

P-Values:

P values can be used to determine whether you FTR or reject H_o. This is the method used in computerized statistical packages. The rule is to reject H_o if $p<\square$. Otherwise, FTR H_o. Let's use our IQ example to demonstrate.

Ex: n = 25 (n < 30) $\rightarrow$ z

σ = 10 (σ known) $\nearrow$

α = .01

$\bar{X}$ = 104

(1) H_o: $\mu \leq 100$

(2) H_A: $\mu > 100$

(2) Instead of using $z_\square$ at this step, we will simply use α=.01. If the p-value generated in step 3 is < than α, then reject H_o.

(3)

$$z^* = \frac{\bar{x} - \mu_o}{\sigma/\sqrt{n}} = \frac{104 - 100}{10/\sqrt{25}} = 2$$

Look up the area associated with the z^* in the normal table. The area =0.4772. Recognize that this is the area from the mean to the right. We want the area out in the right tail, so we must subtract 0.4772 from 0.5, i.e., 0.5-0.4772=.0228. Now we say p=.0228. Draw the picture:

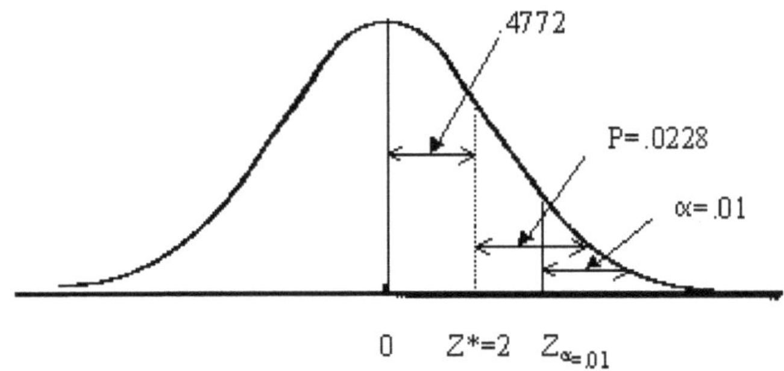

We can see in the picture by comparing the "α" area to the "p" area that z^* would fall to the left of z, and therefore we would FTR H_o.

(3) Fail to reject the H_o. This is the same conclusion we had earlier using z scores.

Just remember the rule: Reject H_o if p<α. Otherwise, FTR H_o. Here p=.0228 and α = 0.01; therefore, p > α, so we FTR H_o.

Confidence interval for population variance σ^2

A Chi-square, χ^2, distribution is used to calculate a CI for σ^2. The χ^2 typically looks like a normal distribution that is skewed to the right, i.e., a long tail to the right.

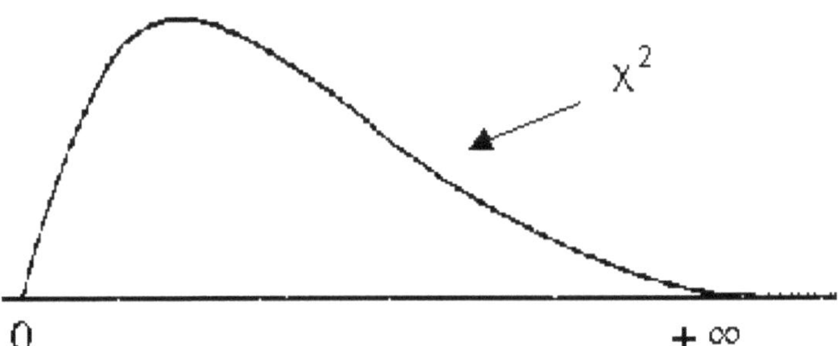

It is a continuous distribution that assumes only positive values. Like the t distribution, the Chi-Square, χ^2 is a 'family' of distributions, where the degrees of freedom determine the shape.

The CI for σ^2 is given by:

$$\frac{(n-1)s^2}{\chi^2_{\alpha/2,\,n-1}} \leq \sigma^2 \leq \frac{(n-1)s^2}{\chi^2_{1-\alpha/2,\,n-1}}$$

Example: Suppose you had the following information:

$\alpha = .05$

$\bar{x}$ $= 5.53$

$\bar{x}$ $n = 10$

$\bar{x}$ $s^2 = .0201$

Calculate a 95% C.I. for σ^2:

$$\frac{(n-1)s^2}{\chi^2_{\alpha/2,\,n-1}} \leq \sigma^2 \leq \frac{(n-1)s^2}{\chi^2_{1-\alpha/2,\,n-1}}$$

$$\frac{(9)(.0201)}{\chi^2_{.025,\,9}} \leq \sigma^2 \leq \frac{(9)(.0201)}{\chi^2_{.975,\,9}}$$

Now, go to the χ^2 table and look up 9 degrees of freedom (df) at $\chi^2_{.025}$ and at $\chi^2_{.975}$. You find 19.0228 and 2.70039, respectively.

$$\frac{(9)(.0201)}{19.0228} \leq \sigma^2 \leq \frac{(9)(.0201)}{2.70039}$$

$$P\,(.0095 \leq \sigma^2 \leq .0670) = .95$$

Hypothesis testing for σ^2

Hypothesis testing for σ^2 is usually of the upper tail type. This is due to the fact that we are usually concerned that a variance is too large.

Our χ^{2*} is calculated:

$X^{2*} = \quad \underline{(n-1)s^2}$

σ_0^2

where σ_0^2 is the variance given in H_0.

Ex: $\square$ = .05, n = 10, $\bar{X}$ = 5.53, s^2 = .0201

1) H_0: $\sigma^2 \leq .2$ $\leftarrow \sigma_0^2$

H_A: $\sigma^2 > .2$

2) Critical value is the table χ^2 value:

$\chi^2_{\alpha,\,n-1} = \chi^2_{.05,\,9} = 16.9190$

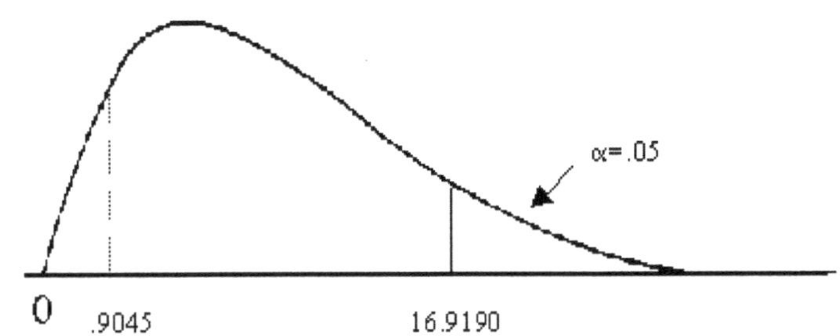

$$\chi^{2*} = \frac{(n-1)s^2}{\sigma^2}$$

3) Test statistics are

4) FTR H_0 at $\alpha = $ $\chi^{2*} = \frac{(n-1)s^2}{\sigma^2} = \frac{9(.0201)}{.2} = .9045$ .05

Now   try   this   example as a two-tail test:

1) H_0: $\sigma^2 = .2$

H_A: $\sigma^2 \neq .2$

2) Critical values are $\chi^2_{\alpha/2,\,n-1}$ (which is $\chi^2_{.025,9}$ = 19.0228) and $\chi^2_{1-\alpha/2,n-1}$ (which = $\chi^2_{.975,9} = 2.70039$)

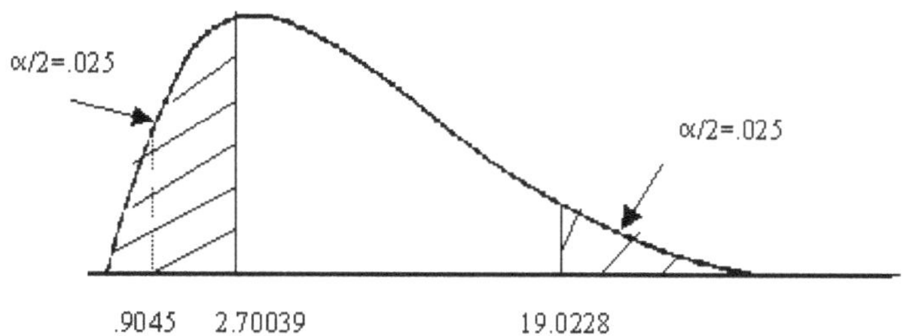

3) Test statistics are

$$= \frac{(9)(.0201)}{.2} = .9045$$

4) Reject H$_o$ at α = .05. Not only is σ² ≠ .2, but it also appears to be < .2 due to the rejection in the left tail.

Chapter Four

Statistical Inference for Means and Variances from Two Populations

C.I. for the Difference in Two Means--Large Independent Samples

Assumptions:

1) Two populations distributed Normally

2) The sample drawn from population 1, n_1, is independent of the sample drawn from population 2, n_2.

3) $n_1 > 30$, $n_2 > 30$

The C.I. is generated by:

$$(\bar{x}_1 - \bar{x}_2) \pm z_{\alpha/2} \sqrt{\frac{\sigma_1^2}{n_1} + \frac{\sigma_2^2}{n_2}}$$

If σ_1^2 and σ_2^2 are unknown, i.e., estimated by s_1^2 and s_2^2, then the C.I. formula is:

$$(\bar{x}_1 - \bar{x}_2) \pm z_{\alpha/2} \sqrt{\frac{s_1^2}{n_1} + \frac{s_2^2}{n_2}}$$

(Note: σ_1^2 is the variance for population 1. σ_2^2 is the variance for population 2. s_1^2 is the variance for sample 1. s_2^2 is the variance for sample 2.)

Ex: An exam was given to 50 seniors at college A and 60 seniors at college B. At A, the mean grade was 75 with a standard deviation of 9. At B, the mean grade was 79 with a standard deviation of 7. Calculate a 95% C.I. for the difference between the performances of the students at A and those at B.

$n_A = 50$, $\bar{x}_A = 75$, $s_A = 9$

$n_B = 60$, $\bar{x}_B = 79$, $s_B = 7$

$\alpha = .05$

$$\left(\overline{x}_A - \overline{x}_B\right) \pm z_{\alpha/2}\sqrt{\frac{s_A^2}{n_A} + \frac{s_B^2}{n_B}}$$

$$= (75 - 79) \pm z_{.025}\sqrt{\frac{9^2}{50} + \frac{7^2}{60}}$$

$$-4 \pm 1.96\sqrt{1.62 + .82}$$

$$-4 \pm 1.96(1.56)$$

$$-4 \pm 3.06$$

$$-7.06, -.94$$

That is to say in other words

$$P(-7.06 \le \mu_A - \mu_B \le -.94) = .95$$

(Note: $\mu_A - \mu_B$ is in the middle of the C.I. and the direction of the subtraction must be the same as you used in the C.I., i.e. $\overline{x}_A - \overline{x}_B \rightarrow \mu_A - \mu_B$ <u>or</u>

$$\overline{x}_B - \overline{x}_A \rightarrow \mu_B - \mu_A)$$

In the example if we had chosen to subtract $\overline{x}_B - \overline{x}_A$ the C.I. would be $P(.94 \le \mu_B - \mu_A \le 7.06) = .95$. Either way you choose to proceed, you can determine from the C.I. that $\mu_B > \mu_A$. This fact is what creates two negative end point values for $\mu_A - \mu_B$ <u>or</u> two positive end point values for $\mu_B - \mu_A$. What if you have one negative and one positive endpoint? We answer through the following example:

Ex: Suppose our C.I. had come out like this: $P(-.94 \le \mu_A - \mu_B \le 7.06) = .95$

Our conclusion would be that μ_A is <u>not</u> different from μ_B. This is because <u>zero</u> is contained in the interval -.94 to 7.06. If zero is contained in the C.I., we <u>cannot</u> say that the difference in the population means is <u>not zero</u>. Hence we accept H_o.

The only way we can conclude that a difference between the means exists is when <u>both</u> end points of the C.I. are either <u>both</u> positive or <u>both</u> negative.

Hypothesis Testing for Two Means--Large, Independent Samples

For this type of test of two means, the test statistic z^* is:

$$z^* = \frac{(\bar{x}_1 - \bar{x}_2) - (\mu_{01} - \mu_{02})}{\sqrt{\dfrac{s_1^2}{n_1} + \dfrac{s_2^2}{n_2}}}$$

Usually, we assume that $(\mu_{01} - \mu_{02}) = 0$. However, in some cases, we might be interested in testing if the difference in the population means some quantity other than zero. In that case, $(\mu_{01} - \mu_{02})$ would be the hypothesized difference. Let's work on the two college examples as a hypothesis test to demonstrate these ideas.

Ex: Is the average grade at college B higher than college A at $\alpha = .05$?

$n_A = 50$, $\bar{X}_A = 75$, $s_A = 9$

$n_B = 60$, $\bar{X}_B = 79$, $s_B = 7$

1) H_0: $\mu_B \leq \mu_A \rightarrow \mu_B - \mu_A \leq 0$

2) H_A: $\mu_B > \mu_A \rightarrow \mu_B - \mu_A > 0$

3) $z_\alpha = z_{.05} = 1.645$ **3)**

$$z^* = \frac{\bar{X}_B - \bar{X}_A}{\sqrt{\dfrac{s_B^2}{n_B} + \dfrac{s_A^2}{n_A}}} = \frac{79 - 75}{\sqrt{\dfrac{7^2}{60} + \dfrac{9^2}{50}}} = \frac{4}{1.56} = 2.56$$

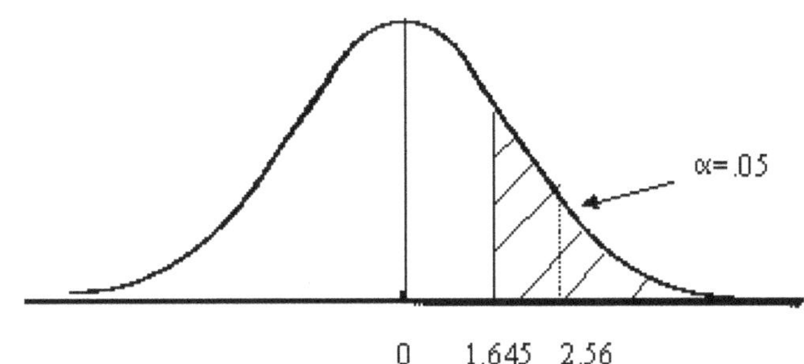

4) Reject H_0 at $\alpha = .05$. The mean grade at college B appears to be greater.

than that at A.

(Note: In this hypothesis test, we assume that $\mu_{oB} - \mu_{oA} = 0$. Therefore, it is not included at the top of the formula for z^*. Also, I chose to subtract $\mu_B - \mu_A$, which yields an upper-tail test. If I had subtracted in the other direction, I would have had a lower-tail test, i.e.

H_o: $\mu_B \leq \mu_A \;\overset{\rightarrow}{}\; \mu_A - \mu_B \geq 0$

H_A: $\mu_B > \mu_A \;\overset{\rightarrow}{}\; \mu_A - \mu_B < 0$

Also, remember α is not divided by 2 in the one-tail test. Our C.I. was two tails, so we had $\alpha/2$.)

Let's try this example with $\mu_{oB} - \mu_{oA} =$ some number.

Ex: Do the students at College B score more than 2 points higher on average than the students at College A. ($\alpha = .05$)?

1) H_o: $\mu_B - \mu_A \leq 2$ $\qquad$ H_A: $\mu_B - \mu_A > 2$

2) $z_\alpha = z_{.05} = 1.645$

3)

$$z^* = \frac{\overline{X}_B - \overline{X}_A - 2}{\sqrt{\dfrac{s_B^2}{n_B} + \dfrac{s_A^2}{n_A}}} = \frac{(79 - 75) - 2}{\sqrt{\dfrac{7^2}{60} + \dfrac{9^2}{50}}} = \frac{2}{1.56} = 1.28$$

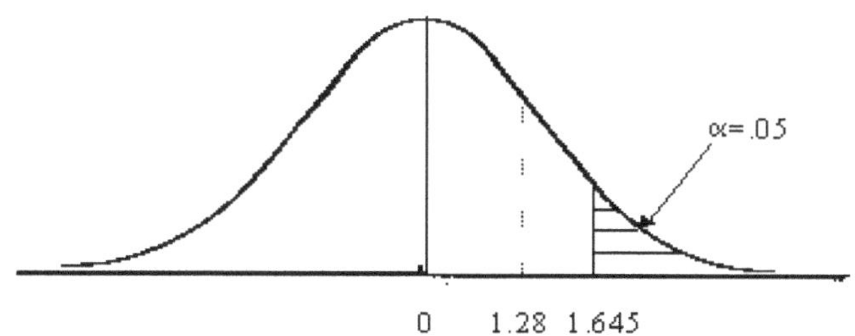

4) FTR H_o at $\alpha = .05$. The students at College B do not score on average 2 points above the students at College A.

Homework (practice problems)

Question B.1

Questions 1 and 2 use almost the same information.

A random sample of 30 households was selected as part of a study on electricity usage, and the number of kilowatt-hours (kWh) was recorded for each household in the sample; the average usage was found to be 375kWh. An extensive study found that the population standard deviation of the usage was given be 81kWh.

Assuming that the user is normally distributed,

a) Formulate and test the hypothesis that the mean household is 400 kWh at a 1% significance level
b) calculate a 99% confidence interval for the mean usage.
c) Use results in (b) to answer (a)

Question B.2

Questions 1 and 2 use *almost* the same information (almost !!!) A random sample of 30 households was selected as part of a study on electricity usage, and the number of kilowatt-hours (kWh) was recorded for each household in the sample for the March quarter of 2006. The average sample usage was found to be 375kWh. If it was found that the sample standard deviation of the usage was 81kWh. Assuming that the usage is normal

a) Formulate and test the hypothesis that the mean household is 400 kWh at a 1% significance level.
b) Calculate a 99% confidence interval for the mean usage.

Use results in (b) to test the hypothesis in (a)

Question B.3

An industrial designer wants to determine the average amount of time it takes an adult to assemble an "easy to assemble" toy. A sample of 16 times yielded an average time of 19.92 minutes, with a sample standard deviation of 5.73 minutes. Assuming normality of assembly times,

(a) provide a 95% confidence interval for the mean assembly time.

(b) Test claim mean is 22 minutes

Top of Form

Question 4

Sample of size n = 100 produced a sample mean of 16. Assuming normality with a population variance $\sigma^2 = 9$, compute a 95% C.I. for population mean μ

Question B.5

What is the smallest sample size required to provide a 95% confidence interval for a mean if it is important that the interval be no longer than 1cm? You may assume that the population is normal with a variance of 9cm.

Question 6

Assuming a normal population with $\sigma^2 = 9$, we want a 95% Confidence Level to estimate the mean μ within a margin of error of how large n (the sample size) should be to achieve that. (A *subtle* challenge to your understanding !!)

Question B.10

The recommended retail price of a brand of designer jeans is $150. The price of the jeans in a sample of 16 retailers is, on average, $141, with a sample standard deviation of 4. If this is a 'random' sample and the prices can be assumed to be normally distributed,

Construct a 95% confidence interval for the average sale price.

Now, do you think that these retailers are satisfied with the recommended retail price or not? Explain why.

Chapter Five

CI and Hypothesis Testing for Two Means

Small Independent Samples with Unequal Variances

(σ_1^2 not equal to σ_2^2)

When $n_1 < 30$ and $n_2 < 30$, we need a 't' distribution with degrees of freedom (df) calculated by:

$$df = \frac{\left(\dfrac{s_1^2}{n_1} + \dfrac{s_2^2}{n_2}\right)^2}{\dfrac{\left(\dfrac{s_1^2}{n_1}\right)^2}{n_1 - 1} + \dfrac{\left(\dfrac{s_2^2}{n_2}\right)^2}{n_2 - 1}}$$

The CI is calculated by:

$$(\bar{x}_1 - \bar{x}_2) \pm t_{\alpha/2,\,df} \sqrt{\frac{s_1^2}{n_1} + \frac{s_2^2}{n_2}}$$

Ex: Suppose you are given the following information and assume that $\sigma_A^2 \neq \sigma_B^2$:

$\bar{x}_A = 118 \qquad \bar{x}_B = 143$

$s_A = 17 \qquad s_B = 24$

$n_A = 9 \qquad n_B = 16$

Calculate a 99% CI for Calculate a 99% CI for $\mu_A - \mu_B$:

$$df = \frac{(\frac{17^2}{9} + \frac{24^2}{16})^2}{\frac{(\frac{17^2}{9})^2}{9-1} + \frac{(\frac{24^2}{16})^2}{16-1}} = \frac{4639.12}{128.89 + 86.4} = 21.55$$

Note:

For the degrees of freedom df to be whole numbers, we will always round down, so df = 21.

The CI is, given by:

$$(\bar{x}_A - \bar{x}_B) \pm t_{.005,21} \sqrt{\frac{s_A^2}{n_A} + \frac{s_B^2}{n_B}}$$

$$= (118 - 143) \pm 2.831 \sqrt{\frac{17^2}{9} + \frac{24^2}{16}}$$

$$= (-25) \pm 2.831(8.25)$$

$$= -25 \pm 23.36$$

$$= -1.64, -48.36$$

Hypothesis $\quad P(-48.36 \leq \mu_A - \mu_B \leq -1.64) = .99 \quad$ testing is done as follows:

The test statistic will be:

$$t^* = \frac{(\bar{x}_1 - \bar{x}_2)}{\sqrt{\frac{s_1^2}{n_1} + \frac{s_2^2}{n_2}}}$$

The df must be calculated as for the CI.

Example/   Let us use the data from the last example and test the following hypothesis at $\alpha = .01$:

1) H_0: $\mu_A = \mu_B \quad \rightarrow \quad \mu_A - \mu_B = 0 \ B = 0$

$\quad H_A$: $\mu_A \neq \mu_B \quad \rightarrow \quad \mu_A - \mu_B \ B \neq 0$

2) $t_{.005,\,21} = \pm\,2.831$

3)
$$t^{*} = \frac{(\bar{x}_A - \bar{x}_B)}{\sqrt{\dfrac{s_A^2}{n_A} + \dfrac{s_B^2}{n_B}}} = \frac{118 - 143}{\sqrt{\dfrac{17^2}{9} + \dfrac{24^2}{16}}} = -3.03$$

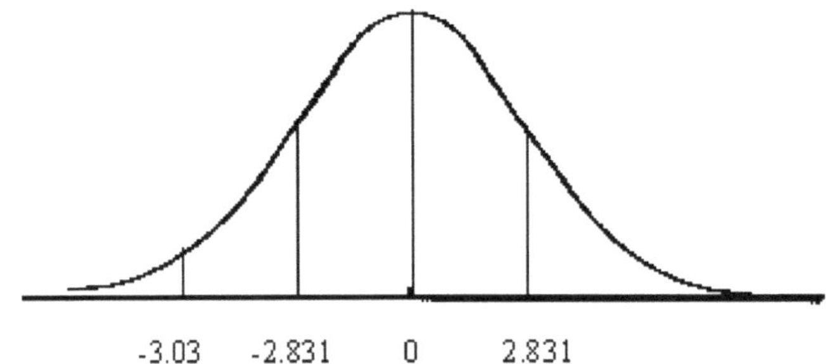

4) Reject H_0 at $\alpha = .01$. $\mu_A \neq \mu_B$. In fact, it appears $\mu_A < \mu_B$.

Small   Independent   Samples   with   Equal   Variance $(\sigma_1^2 = \sigma_2^2)$

(When variances are equal, we can pool the sample variances together)

Now:
$$t^{*} = \frac{\bar{x}_1 - \bar{x}_2}{s_p\sqrt{\dfrac{1}{n_1} + \dfrac{1}{n_2}}}$$

-

Where:
$$s_p = \sqrt{\frac{(n_1 - 1)\,s_1^2 + (n_2 - 1)\,s_2^2}{n_1 + n_2 - 2}}$$

and df $= n_1 + n_2 - 2$. This is known as "pooling." s_p is a weighted average of s_1^2 and s_2^2, where the weights are the associated degrees of freedom.

Example:

Let us try our last example under the assumption $\sigma_A^2 = \sigma_B^2$. (In the next section, we will learn how to test H_0: $\sigma_1^2 = \sigma_2^2$. When we FTR this H_0, then we "pool")

1) $\underline{H_0: \mu_A = \mu_B} \;\overset{\rightarrow}{}\; \underline{\mu_A - \mu_B = 0}$

   $H_A: \mu_A \neq \mu_B \;\overset{\rightarrow}{}\; \mu_A - \mu_B \neq 0$ (two-tailed)

<u>**2)**</u> $t_{\alpha/2,\ nA+nB-2} = t_{.005,\ 23} = \pm 2.807$

<u>**3)**</u>

$$s_p = \sqrt{\frac{(n_A-1)\,s_A^2 + (n_B-1)\,s_B^2}{n_A + n_B - 2}}$$

$$= \sqrt{\frac{(9-1)(17)^2 + (16-1)(24)^2}{9+16-2}} = 21.82$$

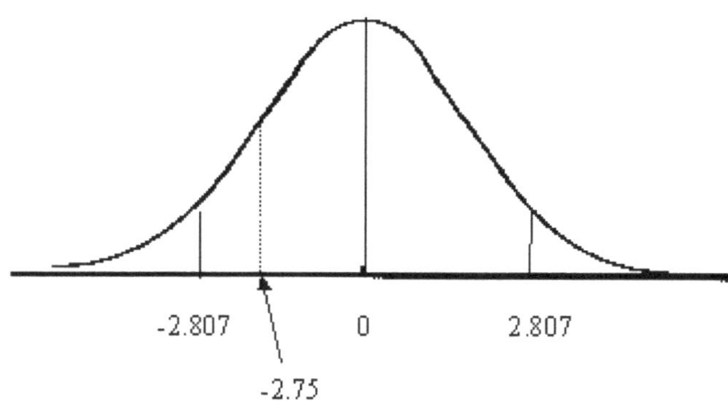

$$t^* = \frac{\overline{x}_A - \overline{x}_B}{s_p\sqrt{\dfrac{1}{n_A} + \dfrac{1}{n_B}}} = \frac{118 - 143}{21.82\sqrt{\dfrac{1}{9} + \dfrac{1}{16}}} = -2.75$$

4) FTR H_o at $\alpha = .01$ $\qquad \mu_A = \mu_B$.

Pooling can really influence your decision. It is important to know when to pool.

That's the next topic.

(Note: Just as with large samples, we may wish to list $\mu_1 - \mu_2 =$ some number. The procedure is the same, i.e., let $\mu_{o1} - \mu_{o2} =$ some number

and if $\sigma_1^2 \neq \sigma_2^2$ use :

$$t^* = \frac{(\bar{x}_1 - \bar{x}_2) - (\mu_{o1} - \mu_{o2})}{\sqrt{\dfrac{s_1^2}{n_1} + \dfrac{s_2^2}{n_2}}}$$

OR

if $\sigma_1^2 = \sigma_2^2$, you pool and:

$$t^* = \frac{(\bar{x}_1 - \bar{x}_2) - (\mu_{o1} - \mu_{o2})}{s_p \sqrt{\dfrac{1}{n_1} + \dfrac{1}{n_2}}}$$

_CI for two means where $\sigma_1^2 = \sigma_2^2$ and $n_1 < 30$ and $n_2 < 30$ are calculated in the usual fashion. Let's illustrate with an example:

Example: A weight loss clinic has an office in Kansas City and one in St. Louis. Both clinics provide exactly the same program for patients. However, the St. Louis clinic prescribes an appetite suppressant for its patients. After 36 weeks, performance is measured. A sample of 15 patients at the KC clinic showed an average weight loss of 78.3 lbs with a standard deviation of 9.2 lbs. A sample of 12 patients at the SL clinic showed a weight loss of 84 lbs with a standard deviation of 8.1 lbs. Calculate a 95% CI for the difference in the patients' performance. Does the appetite suppressant appear to work?

(Assume $\sigma_{kc} = \sigma_{sl}$)

$n_{kc} = 15 \qquad n_{sl} = 12 \quad \alpha = .05$

$\bar{x}_{kc} = 78.3 \qquad \bar{x}_{sl} = 84 \quad S_{kc} = 9.2 \qquad S_{sl} = 8.1$

$$s_p = \sqrt{\frac{(n_{kc} - 1)\, s_{kc}^2 + (n_{sl} - 1)\, s_{sl}^2}{n_{kc} + n_{sl} - 2}}$$

$$= \sqrt{\frac{(15 - 1)(9.2)^2 + (12 - 1)(8.1)^2}{15 + 12 - 2}} = 8.73$$

$$(\bar{x}_{kc} - \bar{x}_{sl}) \pm (t_{\alpha/2, n_{kc}+n_{sl}-2}) * s_p \sqrt{\frac{1}{n_{kc}} + \frac{1}{n_{sl}}}$$

$$= (78.3 - 84) \pm t_{.025,25} * 8.73 \sqrt{\frac{1}{15} + \frac{1}{12}}$$

$$= -5.7 \pm 2.060 \, (3.38)$$

$$= -5.7 \pm 6.96$$

$$= -12.66, 1.26$$

$$= P(-12.66 \leq \mu_{kc} - \mu_{sl} \leq 1.26) = .95$$

No significant difference in weight loss between KC and SL. The appetite suppressant does not appear to work.

Hypothesis Testing Problems

1. A company that packages peanuts states that a maximum of 6% of the peanut shells contain no nuts. At random, 300 peanuts were selected, and 21 of them were empty.

a). With a significance level of 1%, can the statement made by the company be accepted?

b). With the same sample percentage of empty nuts and $1 - \alpha = 0.95$, what sample size would be needed to estimate the proportion of nuts with an error of less than 1%?

2. The life span of 100 W light bulbs manufactured by a particular company follows a normal distribution with a standard deviation of 120 hours, and its half-life is guaranteed under warranty for a minimum of 800 hours. At random, a sample of 50 bulbs from a lot is selected, and it is revealed that the half-life is 750 hours. With a significance level of 0.01, should the lot be rejected for not honoring the warranty?

3. A manufacturer of electric lamps is testing a new production method that will be considered acceptable if the lamps produced by this method result in a normal population with an average life of 2,400 hours and a standard deviation equal to 300. A sample of 100 lamps produced by this method has an average life of 2,320 hours. Can the hypothesis of validity for the new manufacturing process be accepted with a risk equal to or less than 5%?

4. The quality control division of a factory that manufactures batteries suspects defects in the production of a model of mobile phone battery, which results in a lower life for the product. Until now, the time duration in a phone conversation for the battery followed a normal distribution with a mean of 300 minutes and a standard deviation of 30. However, in an inspection of the last batch produced before sending it to market, it was found that the average time spent in conversation was 290 minutes in a sample of 60 batteries. Assuming that the time is still normal with the same standard deviation:

Can it be concluded that the quality control suspicions are true at a significant level of 1%?

5. It is believed that the average level of prothrombin in a normal population is 20 mg/100 ml of blood plasma with a standard deviation of 4 milligrams/100 ml. To verify this, a sample is taken from 40 individuals, of whom the average is 18.5 mg/100 ml. Can the hypothesis be accepted with a significance level of 5%?

Chapter Six

CI and Hypothesis Testing for Two Means Using Paired Samples

Paired samples mean that each subject in the sample participates under treatment one and then under treatment 2. Every subject is used twice. Thus we have two data points for each subject (or subjects may be paired under one treatment (ex: twins))

EXAMPLE/ A company has implemented a new bonus plan. Does the new plan increase sales? The data for a sample of 5 salespeople are as follows:

D I	Before	After	Difference of After − Before
1	15	18	3
2	12	14	2
3	18	19	1
4	15	18	3
5	16	18	2
			---------
			11 total

$$\bar{d} = \frac{\sum d}{n} = \frac{11}{5} = 2.2$$

$$S_d = \sqrt{\frac{\sum d^2 - \frac{(\sum d)^2}{n}}{n-1}} = \sqrt{\frac{\sum (d - \bar{d})^2}{n-1}}$$

Now the four steps of the hypothesis test: (α

$$S_d = \sqrt{\frac{(3-2.2)^2 + (2-2.2)^2 + \ldots + (2-2.2)^2}{5-1}} = \sqrt{\frac{2.8}{4}} = \sqrt{7} = .84$$

= .05)

1) $H_0: \mu_A \leq \mu_B \;\nearrow\; \mu_A - \mu_B \leq 0$ $H_A: \mu_A > \mu_B \;\nearrow\; \mu_A - \mu_B > 0$

2) $t_{\alpha, n-1} = t_{.05, 4} = 2.132$

3)

$$t^* = \frac{\bar{d}}{(s_d/\sqrt{n})} = \frac{2.2}{.84/\sqrt{5}} = \frac{2.2}{.37} = 5.95$$

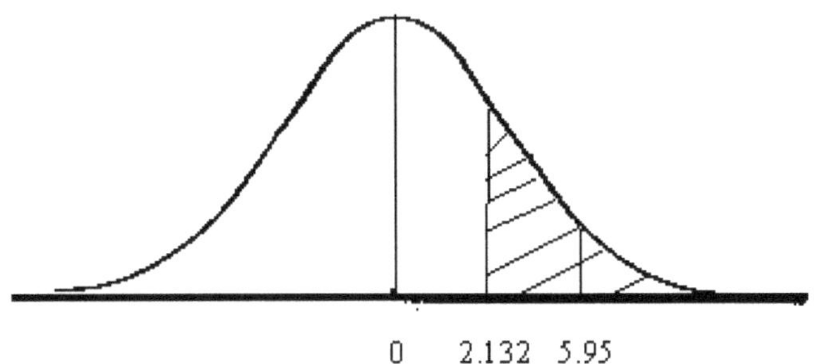

4) Reject H_o at α =0.05. The new bonus plan appears to increase sales.

By how much did sales increase?   Do a 90% CI.

$$\bar{d} \pm t_{\alpha/2,n-1} * \frac{s_d}{\sqrt{n}}$$

$$= 2.2 \pm t_{.05,4} * \frac{.84}{\sqrt{5}} = 2.2 \pm 2.132\,(.37)$$

$$= 1.41,\, 2.99$$

$$P\,(1.41 \le \mu_A - \mu_B \le 2.99) = .90$$

CI and Hypothesis Testing for Two Variances

Test of two variances, σ_1^2 and σ_2^2, require an <u>F distribution</u>.

The F distribution:

1) is continuous

2) is a "family" of distributions where a <u>pair</u> of <u>df</u> determines its shape

3) is typically skewed right like σ^2

4) takes on only <u>positive</u> values (like the Chi2 distribution)

Let's use three examples to illustrate one-tail and two-tail hypothesis tests and a two-tail CI.

EXAMPLE (1)/   We have two sources of raw material, A & B. Required to test: Is the variance in source A greater than B, at $\alpha= .01$?

$n_A = 10$ $n_B = 11$ $s_A^2 = 250$ $s_B^2 = 195$

1) $H_0: \sigma_A^2 \leq \sigma_B^2$ $H_A: \sigma_A^2 > \sigma_B^2$

2) $F_{\alpha, n_A - 1, n_B - 1} = F_{.01, 9, 10} = 4.94$

(Note: The pair of degrees of freedom must be in the order: <u>larger variance's degrees of freedom first, while smaller variance's df second</u>. Here $s_A^2 = 250 > s_B^2 = 195$, thus $F_{\alpha, n_A - 1, n_B - 1}$)

(Note: In order to find $F_{.01, 9, 10}$:

1)   Find the .01   F-Table.

2)   Look up numerator df 9 and denominator df 10. The df is in this order.

because $s_A^2 > s_B^2$, and in step 3, our

$$F^* = \frac{s_A^2}{s_B^2}$$

$\leftarrow$ Numerator

$\leftarrow$ Denominator

The rule in hypothesis testing of two variances: <u>Always put the larger sample variance at the top of the F^* fraction</u>. This rule allows us to only have work with the upper tail of the F distribution, which are the only values tabulated)

3)

$$F^* = \frac{s_A^2}{s_B^2} = \frac{250}{195} = 1.282$$

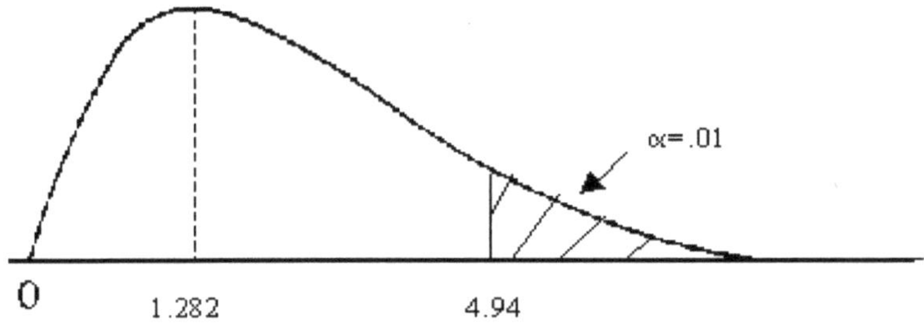

4) FTR H$_0$ at α = .01.   i.e.   Variance of A $\leq$ Variance of B

Let's work on this same problem as a two-tail hypothesis test at <u>α = .05</u>.

(Notice: This text has F tables for α = .10, .05, .025, and .01)

1)   H$_0$: $\sigma_A^2 = \sigma_B^2$

2)   H$_A$: $\sigma_A^2 \neq \sigma_B^2$

2) Critical value of interest is the <u>upper tail only</u> because F* is constructed larger variance over a smaller variance.

$F_{\alpha/2,\, n_A-1,\, n_B-1} = F_{.025,\, 9,\, 10} = 3.78$

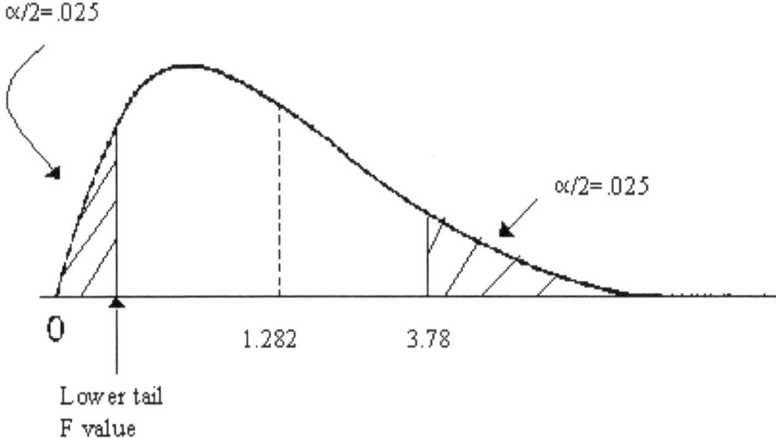

3) $F^* = \dfrac{s_A^2}{s_B^2} = \dfrac{250}{195} = 1.282$

Remember, since F* is constructed with larger variance over smaller variance, the only possibility of rejection is in the upper tail (even though this is a two-tail test). Therefore, we don't need to worry about looking up the lower tail critical value.

4) FTR H_o at $\alpha = .05$ Variance of A = variance of B

Now do a 95% CI for σ_A^2 / σ_B^2

The CI for σ_A^2 / σ_B^2 is :

$$\frac{\dfrac{s_A^2}{s_B^2}}{F_{\alpha/2,\,n_A-1,\,n_B-1}} \leq \sigma_A^2 / \sigma_B^2 \leq \frac{\dfrac{s_A^2}{s_B^2}}{\left(\dfrac{1}{F_{\alpha/2,\,n_B-1,\,n_A-1}}\right)}$$

This is known as the reciprocal property of the F distribution. The df's are reversed here.   (for getting the lower tail F value)

In our example

$$\frac{s_A^2}{s_B^2} = \frac{250}{195} = 1.282$$

$$F_{\alpha/2,\,n_A-1,\,n_B-1} = F_{.025,9,10} = 3.78$$

$$\frac{1}{F_{\alpha/2,\,n_B-1,\,n_A-1}} = \frac{1}{F_{.025,10,9}} = \frac{1}{3.96} = .25$$

$$P\left(\frac{1.282}{3.78} \leq \sigma_A^2 / \sigma_B^2 \leq \frac{1.282}{.25}\right) = .95$$

$$P\left(.34 \leq \sigma_A^2 / \sigma_B^2 \leq 5.128\right) = .95$$

The 95% CI for the variance ratio σ_A / σ_B is: **(.34, 5.128)**

Usually, we would interpret a CI for the ratio of two variances as σ_A^2 is .34 (the lower confidence limit) to 5.128 (the upper confidence limit) <u>times larger than</u> σ_B^2. However, because one is contained in this CI, we can conclude that $\sigma_A^2 = \sigma_B^2$.

CI and Hypothesis Testing for Two Proportions

Assumptions:

1) $n_1 \geq 30$ and $n_2 \geq 30$

2) $n_1 \hat{p}_1 > 5$ and $n_2 \hat{p}_2 > 5$
   $n_1(1-\hat{p}_1) > 5$ and $n_2(1-\hat{p}_2) > 5$

(Note: $p_1 = x_1/n_1$ where x_1=observed number of "successes" in sample 1 and similarly $p_2 = x_2/n_2$ where x_2=observed number of "successes" in sample 2.)

The CI is:

$$(\hat{p}_1 - \hat{p}_2) \pm Z_{\alpha/2} \sqrt{\frac{\hat{p}_1(1-\hat{p}_1)}{n_1} + \frac{\hat{p}_2(1-\hat{p}_2)}{n_2}}$$

EXAMPLE

A pharmaceutical company is testing two new drugs intended to reduce blood pressure. Two groups of volunteers are used. In group 1, 71 of 100 patients tested responded to drug 1 with lower blood pressure. In group 2, 58 of 90 patients tested responded to drug 2 with lower blood pressure. Calculate a 95% CI for the difference in the efficacies of the two drugs.

$$n_1 = 100 \qquad n_2 = 90 \qquad \alpha = .05$$
$$\hat{p}_1 = 71/100 = .71 \qquad \hat{p}_2 = 58/90 = .64$$

Then:

$$(\hat{p}_1 - \hat{p}_2) \pm z_{\alpha/2} \sqrt{\frac{\hat{p}_1(1-\hat{p}_1)}{n_1} + \frac{\hat{p}_2(1-\hat{p}_2)}{n_2}}$$

$$= (.71 - .64) \pm 1.96 \sqrt{\frac{.71(1-.71)}{100} + \frac{.64(1-.64)}{90}}$$

$$= .07 \pm 1.96(.07) = -.07 , .21$$

$$P(-.07 \le p_1 - p_2 \le .21) = .95$$

No difference in the efficacy of the two drugs because C.I. includes 0

<u>Hypothesis tests</u> of two proportions use a "pooled" estimate of the proportion because under H_0 $p_1 = p_2$ (leading to equality of variances.). The pooled estimate of the proportions is:

$$\bar{p} = \frac{x_1 + x_2}{n_1 + n_2}$$

And the standard error of this estimate becomes:

$$\sqrt{\frac{\bar{p}(1-\bar{p})}{n_1} + \frac{\bar{p}(1-\bar{p})}{n_2}}$$

and

$$z^* = \frac{\hat{p}_1 - \hat{p}_2}{\sqrt{\frac{\bar{p}(1-\bar{p})}{n_1} + \frac{\bar{p}(1-\bar{p})}{n_2}}}$$

Let's redo our last example as a hypothesis test:

EXAMPLE/   Is there a significant difference between the efficacies of the two drugs at $\alpha=.05$?

$n_1 = 100$ $n_2 = 90$ (large sample sizes) $x_1 = 71$ $x_2 = 58$

1) H_o: $p_1 = p_2$ $\quad\quad$ -----> $\quad\quad$ $p_1 - p_2 = 0$

H_A: $p_1 \neq p_2$ $\quad$ ----> $\quad$ $p_1 - p_2 \neq 0$ $\quad\quad$ (two-tailed test)

2) $\pm Z_{\alpha/2} = \pm Z_{.025} = \pm 1.96$

3)
$$\bar{p} = \frac{71+58}{100+90} = .68$$

$$z^* = \frac{\hat{p}_1 - \hat{p}_2}{\sqrt{\dfrac{\bar{p}(1-\bar{p})}{n_1} + \dfrac{\bar{p}(1-\bar{p})}{n_2}}} = \frac{\dfrac{71}{100} - \dfrac{58}{90}}{\sqrt{\dfrac{.68(1-.68)}{100} + \dfrac{.68(1-.68)}{90}}}$$

$$= \frac{.71 - .64}{.07} = \frac{.07}{.07} = 1.0$$

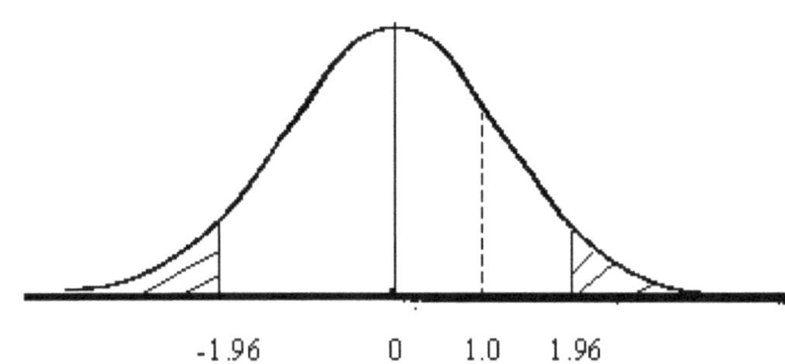

4) FTR H_o at $\alpha = .05$. The drugs appear to have the same efficacy.

Japanese Girls' Names by Kumi Furuichi It used to be very typical for Japanese girls' names to end with "ko." (The trend might have started around my grandmothers' generation and its peak might have been around my mother's generation.) "Ko" means "child" in Chinese character. Parents would name their daughters with "ko" attaching to other Chinese characters which have meanings that they want their daughters to become, such as Sachiko – a happy child, Yoshiko – a good child, Yasuko – a healthy child, and so on. However, I noticed recently that only two out of nine of my Japanese girlfriends at this school have names which end with "ko." More and more, parents seem to have become creative, modernized, and, sometimes, westernized in naming their children. Chapter 9 Source URL: http://cnx.org/content/m17001/latest/ Saylor URL: http://saylor.org/courses/bus204 Attributed to: [Susan Dean and Barbara Illowsky] Saylor.org Page

 I have a feeling that, while 70 percent or more of my mother's generation would have names with "ko" at the end, the proportion has dropped among my peers. I wrote down all my Japanese friends', ex-classmates', co-workers, and acquaintances' names that I could remember. Below are the names. (Some are repeats.) Test to see if the proportion has dropped for this generation. Data: Ai, Akemi, Akiko, Ayumi, Chiaki, Chie, Eiko, Eri, Eriko, Fumiko, Harumi, Hitomi, Hiroko, Hiroko, Hidemi, Hisako, Hinako, Izumi, Izumi, Junko, Junko, Kana, Kanako, Kanayo, Kayo, Kayoko, Kazumi, Keiko, Keiko, Kei, Kumi, Kumiko, Kyoko, Kyoko, Madoka, Maho, Mai, Maiko, Maki, Miki, Miki, Mikiko, Mina, Minako, Miyako, Momoko, Nana, Naoko, Naoko, Naoko, Noriko, Rieko, Rika, Rika, Rumiko, Rei, Reiko, Reiko, Sachiko, Sachiko, Sachiyo, Saki, Sayaka, Sayoko, Sayuri, Seiko, Shiho, Shizuka, Sumiko, Takako, Takako, Tomoe, Tomoe, Tomoko, Touko, Yasuko, Yasuko, Yasuyo, Yoko, Yoko, Yoko, Yoshiko, Yoshiko, Yoshiko, Yuka, Yuki, Yuki, Yukiko, Yuko, Yuko.EXERCISE 7 An article in the San Jose Mercury News stated that students in the California state university system take an average of 4.5 years to finish their undergraduate degrees. Suppose you believe that the average time is longer.

Homework (practice problems)

Question B.1

Questions 1 and 2 use almost the same information

A random sample of 30 households was selected as part of a study on electricity usage, and the number of kilowatt-hours (kWh) was recorded for each household in the sample; the average usage was found to be 375kWh. In a very large study, it was found that the population standard deviation of the usage was gn to be 81kWh.

Assuming that the usage is normally distributed,

Formulate and test the hypothesis that the mean household is 400 kWh at a 1% significance level.

calculate a 99% confidence interval for the mean usage.

Use results in (b) to answer (a)

Question B.2

Questions 1 and 2 use _almost_ the same information (almost !!!)

A random sample of 30 households was selected as part of a study on electricity usage, and the number of kilowatt-hours (kWh) was recorded for each household in the sample for the March quarter of 2006. The average sample usage was found to be 375kWh. If it was found that the sample standard deviation of the usage was 81kWh. Assuming that the usage is normal

, Formulate and test the hypothesis that the mean household is 400 kWh at a 1% significance level.

calculate a 99% confidence interval for the mean usage.

Use results in (b) to test the hypothesis in (a)

Question B.3

An industrial designer wants to determine the average amount of time it takes an adult to assemble an "easy to assemble" toy. A sample of 16 times yielded an average time of 19.92 minutes, with a sample standard deviation of 5.73 minutes. Assuming normality of assembly times,

(a) provide a 95% confidence interval for the mean assembly time.

(b) Test claim mean is 22 minutes

Question 4

Sample of size n = 100 produced a sample mean of 16. Assuming normality with a population variance $\sigma^2 = 9$, compute a 95% C.I. for population mean μ

Question B.5

What is the smallest sample size required to provide a 95% confidence interval for a mean if it is important that the interval be no longer than 1cm? You may assume that the population is normal with a variance of 9cm.

Chapter Seven

Analysis Of Variance (ANOVA)

Definition: **Analysis of variance** (**ANOVA**) is a collection of statistical models used to analyze the differences between group means and their associated procedures (such as "variation" among and between groups); and was developed by Sir **Ronald A. Fisher** (1890 – 1962) who was an English statistician, evolutionary biologist, geneticist, and eugenicist who is also responsible for developing Design of Statistical Experiments.

With ANOVA we are testing:

H_0: $\mu_1 = \mu_2 = \mu_3 = ... = \mu_m$

H_A: Not all the μ_j's are equal

(Note: H_A is <u>not</u> written $\mu_1 \neq \mu_2 \neq \mu_3 \neq ... \neq \mu_m$. ***Why do you think? Because we can have at least one*** μ_j that is different from the rest and is responsible for increasing the variance)

Simply stated, ANOVA is a test of more than two means.

Simplified Example: We are studying consumer preference for the color of a new soft drink. The colors are (1) colorless, (2) pink, (3) orange, and (4) green. We randomly assign each color to 5 grocery stores. At the end of one month, we record the number of cases sold per 1,000 population in the store's market area. We want to determine which color has the highest average sales. (Remember, the product is identical except for color. Color is our treatment – in different levels: red, yellow, green, and orange) (Note: treatment can be a real treatment as in medicine, fertilizer, or any Factor)

Assumptions:

1) Sales are normally distributed.

2) Each normal distribution has the same variance, σ^2.

3) Only the means (**μ_j**) of the normal distributions may differ.

A picture might look as follows:

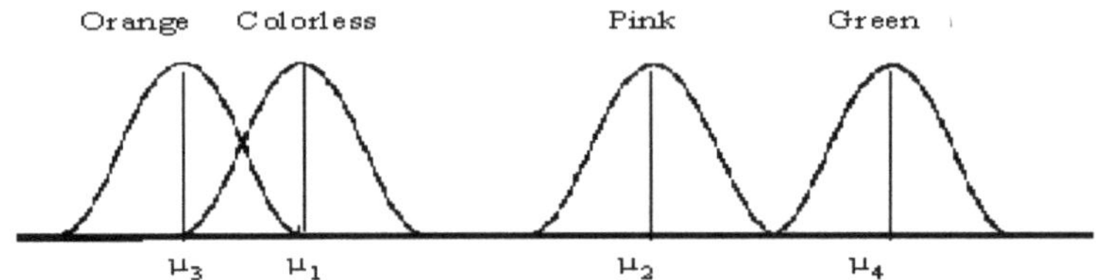

(All of the normal curves look identical <u>because</u> all have the same σ^2. Only the location of the distributions on the number line may differ, i.e., not all μ_j's may be equal.)

Suppose the data look as follows:

Treatment j

Obs.	1	2	3	4
I	Colorless	Pink	Orange	Green
1	26.5	31.2	27.9	30.8
2	28.7	28.3	25.1	29.6
3	25.1	30.8	28.5	32.4
4	29.1	27.9	24.2	31.7
5	27.2	29.6	26.5	32.8
Total	136.6	147.8	132.2	157.3

<u>Point Estimation of μ_j</u>

$\overline{Y}_j$ Is the point estimate of the treatment mean μ_j. (Remember, treatment here is color.)

$$\overline{Y}_j = \frac{\sum_{i-1}^{n_j} Y_{ij}}{n_j}$$

EX/ $\overline{Y}_1$ = point of the estimate of μ_1

$$= \frac{\sum_{i-1}^{n_1} Y_{i1}}{n_1}$$

$$= \frac{\sum_{i-1}^{5} Y_{i1}}{5}$$

$$= \frac{26.5 + 28.7 + 25.1 + 29.1 + 27.2}{5}$$

$$= \frac{136.6}{5} = 27.32$$

=the sample mean for treatment 1, colorless

And

$\overline{Y}_2$ = point estimate of μ_2m = the sample mean for treatment 2, pink

$$= \frac{\sum_{i=1}^{n_2} Y_{i2}}{n_2} = \frac{\sum_{i=1}^{5} Y_{i2}}{5}$$

$$= \frac{31.2 + 28.3 + 30.8 + 27.9 + 29.6}{5}$$

$$= \frac{147.8}{5} = 29.56$$

Similarly;

$\overline{Y}_3 = 26.44$ and

$\overline{Y}_4 = 31.46$

(Note: The highest mean sales is for treatment 4, green, $\overline{Y}_4=31.46$. The next highest is for treatment 2, pink, $\overline{Y}_2=29.56$.)

Partitioning The Total Sum Of Squares (SSTO)

Steps:

1) Compute an overall sample mean (grand mean).

$$\overline{\overline{Y}} = \frac{\sum_i \sum_j Y_{ij}}{n_T} \qquad \text{where } n_T = \sum_j n_j$$

(Y_{ij} = individual observation and n_T = total sample size)

EXAMPLE:

Add up all observations and divide by the total number of observations.

$$\overline{\overline{Y}} = \frac{[(26.5+28.7+25.1+29.1+27.2)+(31.2+...+29.6)}{(5+5+5+5+5)}$$
$$\frac{+(27.9+...+26.5)+(30.8+...+32.8)]}{}$$

$$= \frac{573.9}{20} = 28.695$$

2) Compute SSTO = the deviations of the individual observations, Y_{ij}, from the overall mean, $\overline{\overline{Y}}$, squared and summed

EX/ $\quad SSTO = \sum_j \sum_i (Y_{ij} - \overline{\overline{Y}})^2$

SSTO = $(26.5-28.695)2 + (28.7-28.695)2 + ... + (32.8-28.695)2 = 115.92950$ = take each observation to subtract the overall mean from it and square the   Quantity, add them all up.

3) Compute the Sum of Squares for Treatment (SSTR), which is the treatment means, $\overline{Y}_j$, minus the overall mean, square the quantities, weigh them by the number of observations in the treatment group, n_j, and sum them.

$$SSTR = \sum_{j=1}^{4} n_j (\overline{Y}_j - 28.695)^2$$

$$= 5(27.32 - 28.695)^2 + 5(29.56 - 28.695)^2 + 5(26.44 - 28.695)^2$$

$$+ 5(31.46 - 28.695)^2$$

$$SSTR = \sum_{j} n_j (\overline{Y}_j - \overline{\overline{Y}})^2 \qquad = 76.84550$$

4) Compute the Sum of Squares for Error (SSE), which is the individual observations, Y_{ij}, minus its treatment mean, $\overline{Y}_j$, squared, then sum.

$$SSE = \sum_{j} \left[\sum_{i} (Y_{ij} - \overline{Y}_j)^2 \right]$$

$$SSE = \sum_{j=1}^{4} \left[\sum_{i=1}^{5} (Y_{ij} - \overline{Y}_j)^2 \right]$$

$$= [(26.5 - 27.32)^2 + ... + (27.2 - 27.32)^2] + [(31.2 - 29.56)^2$$

$$+ ... + (29.6 - 29.56)^2]$$

$$+ [(27.9 - 26.44)^2 + ... + (26.5 - 26.44)^2] + [(30.8 - 31.46)^2$$

$$+ ... + (32.8 - 31.46)^2]$$

$$= 39.08400$$

Now that we know the mechanics of how SSTO, SSTR, and SSE are computed, how do they fit together?

SSTO = SSTR + SSE

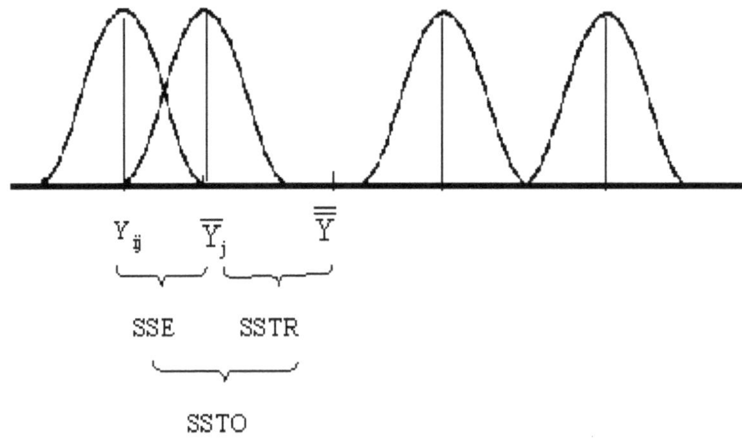

EX/ $\quad$ SSTO = SSTR + SSE

$$115.92950 = 76.84550 + 37.08400$$

This is known as partitioning the total sum of squares into its parts - sum of squares for treatment and sum of squares for error. (Note: If the treatment means $\bar{Y}_j$'s are close to each other, SSTR will be very close to zero. SSTR is very large when the treatment means $\bar{Y}_j$'s are <u>not</u> close to each other.)

Partitioning The Degrees Of Freedom

We know that variance is a sum of squares divided by degrees of freedom. We want to turn SSTR and SSE into variances by dividing them by their df. First, we have to calculate the df that goes with SSTO, SSTR, and SSE.

1) SSTO has $n_T - 1$ df because there are n_T deviations $(Y_{ij} - \bar{\bar{Y}})$ but one

Constraint exists, namely $\sum \sum (Y_{ij} - \bar{\bar{Y}}) = 0$. This means that when you sum

Up You must get zero for all of the deviations (not squared) remember, some of the deviations $(Y_{ij} - \bar{\bar{Y}})$ will be positive, and others will be negative or zero.

2) SSTR has r-1 df where r is the number of treatment groups that you have (j = 1, 2, …, r) SSTR has r-1 df because there are r deviations ($\overline{Y}_j - \overline{\overline{Y}}$), but one constraint exists, namely $\sum_j n_j (\overline{Y}_j - \overline{\overline{Y}}) = 0$.

The logic behind this constraint is exactly the same as we just discussed for SSTO.

3) SSE has $n_T - r$ df because the r trt groups each contribute a component SS, $\sum_i^{nj} \sum_j^r (Y_{ij} - \overline{Y}_j)^2$ to SSE. The component SS for the jth treatment is equivalent in form to a total SS and therefore has $n_j - 1$ df. The df associated with SSE is then the sum of the df for each of the r components:

$$\sum_{j=1}^r (n_j - 1) = \left(\sum_{j=1}^r n_j\right) - r = n_T - r$$

<u>Note</u>: Most people get the df for SSE by subtraction. This is possible because df are addictive, just like the SS:

$n_T - 1 = r - 1 + n_T - r$

$\uparrow \qquad \uparrow \qquad \uparrow$

df for   df for   df for

SSTO   SSTR   SSE

EX/   (20-1) = (4-1) + (20-4)

     19   =   3   +   16

     $\uparrow$ $\uparrow$ $\uparrow$

       df for   df for   df for

      SSTO   SSTR   SSE

Mean Squares (MS)

Mean square is another name for variance. A mean square is a SS divided by df. There are two mean squares that are of interest to us:

$$\textbf{MSTR = treatment mean square} = \frac{SSTr}{r-1}$$

$$MSTR = \frac{76.84550}{3} = 25.61517$$

MSE = error mean square $\quad = \dfrac{SSE}{n_T - r}$

In the example: $\quad \text{MSE} = \dfrac{39.08400}{16} = 2.44275$

F Test for Equality of Treatment Means

Recall the hypothesis that we want to test:

H_0: $\mu_1 = \mu_2 = \ldots = \mu_r$

H_A: Not all μ_j's are equal

Our test statistic will be the ratio of the two mean squares (variances), which we know will be distributed F. The test statistic is

$$F^* = \dfrac{\text{MSTR}}{\text{MSE}}$$

A large value for F^* leads us to reject H_0 because of MSTR > MSE. If F^* is

Approximately equal to 1, then MSTR $\approx$ MSE, which leads us to FTR H_0.

(Note: It is a statistical property that MSTR $\geq$ MSE. Therefore, it is customary in ANOVA to place the larger variance MSTR at the top of the F^* fraction. Therefore, our F test is always of the upper tail type.)

The F test is conducted as usual where the df are (r-1, n_T -r)

EX/ H_0: $\mu_1 = \mu_2 = \mu_3 = \mu_4$ H_A: not all μ_j's are equal

$\alpha = .05$ r - 1 = 3 n_T - r = 16

$F_{.05,\,3,\,16} = 3.24$ (F table value)

$$F^* = \frac{MSTR}{MSE} = \frac{25.61517}{2.44275} = 10.49$$

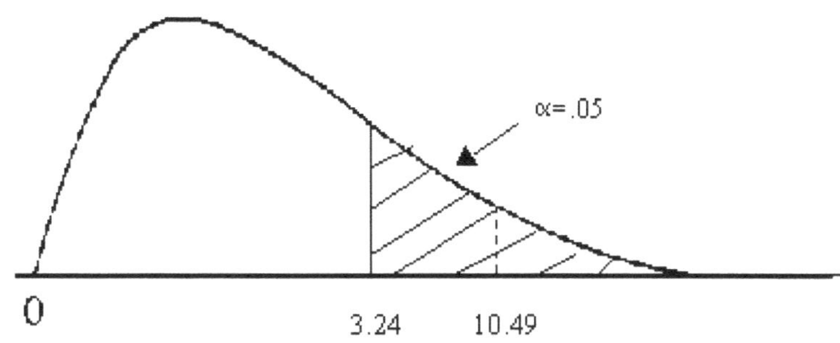

Reject H_o. Not all colors sell equally well.

ANOVA Table

An ANOVA table is a summary of the analysis. The general format is as follows:

Source	d.f.	SS	MS	F
TRT	$r - 1$	SSTR	MSTR	F^*
ERROR	$n_T - r$	SSE	MSE	
TOTAL	$n_T - 1$	SSTO		

For the example above, we have the following:

Source	d.f.	SS	MS	F
TRT	3	76.84550	25.61517	10.49
ERROR	16	39.08400	2.44275	
TOTAL	19	115.92950		

Analysis of specific individual Treatment Effects

A) Simple Pairing Methods:

When we <u>reject H_o</u> and conclude that all of the $\square_j$'s are not equal, we need to

Analyze for treatment effects, i.e., where do the differences lie? We will discuss two procedures: 1) estimation of the mean response and 2) comparison of mean responses for two treatments. These are equivalent to a CI for a single mean and a CI for the difference between two means.

(<u>Note</u>: There are <u>many</u> methods for analyzing treatment effects. The above are only two of them.)

1) <u>Estimation of μ_j</u>

$\overline{Y}_j$ is our point estimate of μ_j.

and the estimated variance of $\overline{Y}_j$ is.

$$s_{\overline{Y}_j}^2 = \frac{MSE}{n_j}$$

(<u>Note</u>: Recall when you used to estimate μ, $\sqrt{s^2/n}$ was the standard error. That is exactly what is going on in ANOVA; only the symbols are different. Here you use to estimate μ_j using $\overline{Y}_j$, and $\sqrt{MSE/n_j}$ because you have more than one normal distribution, hence the subscript j)

The CI for μ_j is

$$\overline{Y}_j - t_{n_T - r}\sqrt{\frac{MSE}{n_j}} \leq \mu_j \leq \overline{Y}_j + t_{n_T - r}\sqrt{\frac{MSE}{n_j}}$$

<u>Note</u>: The t has $(n_T - r)$ df because this is the df associated with MSE.

In the example/ we could compute a CI for each of the four μ_j's.

Let's calculate a 90% CI for μ_1 to illustrate the procedure:

$\overline{Y}_1 = 27.32$ MSE = 2.44275

$n_1 = 5$ $n_T - r = 16$

$$\overline{Y}_1 \pm t_{.05,16}\sqrt{\frac{MSE}{n_j}}$$

$$27.32 \pm 1.746\,(.69896)$$

$$P\,(\,26.1 \leq \mu_1 \leq 28.5\,) = .90$$

With 90% confidence, mean sales for the colorless version are between 26.1 and 28.5 cases per 1,000 population. (You do the CI's for μ_2, μ_3, and μ_4 for practice.)

2) <u>Comparison of Mean Response for Two Treatments</u>

Suppose we want to compare the mean responses for two treatments., j and j'. We are interested in a CI of the difference between the means, μ_j, and $\mu_{j'}$. The point estimator of this difference is $\overline{Y}_j$ - $\overline{Y}_{j'}$. The variance associated with this difference is the sum of the individual variances:

$$s^2_{\overline{Y}_j-\overline{Y}_{j'}} = \frac{MSE}{n_j} + \frac{MSE}{n_{j'}}$$

The CI is:

$$(\overline{Y}_j - \overline{Y}_{j'}) - t_{n_T-r}\sqrt{\frac{MSE}{n_j} + \frac{MSE}{n_{j'}}}$$

$$\leq \mu_j - \mu_{j'} \leq$$

$$(\overline{Y}_j - \overline{Y}_{j'}) + t_{n_T-r}\sqrt{\frac{MSE}{n_j} + \frac{MSE}{n_{j'}}}$$

In the example/ 	Suppose we want to compare the effects of green and pink. The pairwise comparison is μ_4 - μ_2. Calculate a 90% CI.

$$\overline{Y_4} - \overline{Y_2} = 31.46 - 29.56 = 1.90$$

$$t_{.05,16} = 1.746$$

$$s_{\overline{Y_4}-\overline{Y_2}} = \sqrt{\frac{MSE}{n_4} + \frac{MSE}{n_2}}$$

$$= \sqrt{\frac{2.44275}{5} + \frac{2.44275}{5}}$$

$$= \sqrt{.97710} = .98848$$

$$(\overline{Y_4} - \overline{Y_2}) \pm t_{.05,16}\, s_{\overline{Y_4}-\overline{Y_2}}$$

$$1.90 \pm 1.746\,(.98848) = .2, 3.6$$

$$P(.2 \leq \mu_4 - \mu_2 \leq 3.6) = .90$$

With 90% confidence, mean sales for green are between .2 and 3.6 cases per 1,000 population, higher than those for pink. (You make the other paired comparisons for practice, i.e., one vs2,   2 vs. 3,   3 vs. 4,   etc.)

B) Tukey Testing Method:

The Studentized Range (q) is the difference between the largest and smallest data point in a sample, measured in terms of sample standard deviations. The studentized range distribution is the probability distribution of studentized ranges for independent, identically distributed random variables that are normally distributed. It is primarily used in post hoc tests, like Tukey's HSD, to limit the Type I error risk.

The shape of the studentized range distribution depends upon the context. For example, if you are testing whether two means are equal, it is similar to the T distribution. However, it takes the number of means into account. The more means, the larger the critical value.

An ANOVA test can tell you if your results are significant overall, but it will not tell you exactly where those differences lie. After you have run an ANOVA and found significant results, then you can run Tukey's HSD to find out which specific groups' means (compared with each other) are different. The test compares all possible pairs of means.

Formula:

The formulas differ a little depending on what post hoc test you are running. For example, the formula for calculating a q ratio for differences between means is:

To test all pairwise comparisons among means using the Tukey HSD, calculate HSD for each pair of means using the following formula:

$$HSD = \frac{M_i - M_j}{\sqrt{\dfrac{MS_w}{n_h}}}$$

Where:

- $M_i - M_j$ is the difference between the pair of means., M_i should be larger than M_j
- MS_w is the Mean Square Within, sometimes denoted by MSE
- n is the sample number in the group or treatment.

Like most hypothesis tests, a calculated value (from the formula) is compared with a table value (from the q critical value table below). In order to use this particular formula, you would have to calculate a q ratio for each pair of means to see if there is a statistically significant difference for each pair. An alternate formula (which is really just a rearrangement of the formula above) is based on a difference between means instead of a q ratio:

$$HSD = q_{critical} \times \sqrt{\frac{MS_{within}}{n}}$$

For more details on how to run a post hoc test using the studentized range distribution, see:

<u>Tukey's HSD.</u>

<u>Duncan's Multiple Range Test.</u>

<u>Newman Keuls.</u>

General Tukey Test Steps

Step 1: Perform the **ANOVA test**. Assuming your F value is significant, you can run the post hoc test.

Step 2: Choose two means from the ANOVA output. Note the following:

- Means,
- **Mean Square Within** (MSE)
- n   Sample number per treatment/group,
- Degrees of freedom of Within-sample. (i.e., Error d. of f.)

Step 3: Calculate the HSD statistic for the Tukey test using the formula.

Step 4: Find the score in **Tukey's critical value table**.

Step 5: Compare the score you calculated in Step 3 with the tabulated value you found in Step 4. If the calculated value from Step 3 is bigger than the critical value from the critical value table, the two means are <u>significantly different</u>.

Assumptions for the test

- Observations are independent within and among groups.
- The groups for each mean in the test are <u>normally distributed</u>.
- There is equal within-group variance across the groups associated with each mean in the test (<u>homogeneity of variance</u>).

References:

1) Duncan, David B. [1955]. Multiple ranges and multiple F tests. Biometrics li, 1-42.
2) H. Leon Harter. Critical Values for Duncan's New Multiple Range Test. Biometrics, Vol. 16, No. 4,
3) (Dec. 1960), pp. 671-685

Rebecca Warner. <u>Applied Statistics.</u>

	2	3	4	5	6	7	8	9	10
5	3.64	4.60	5.22	5.67	6.03	6.33	6.58	6.80	6.99
6	3.46	4.34	4.90	5.30	5.63	5.90	6.12	6.32	6.49
7	3.34	4.16	4.68	5.06	5.36	5.61	5.82	6.00	6.16
8	3.26	4.04	4.53	4.89	5.17	5.40	5.60	5.77	5.92
9	3.20	3.95	4.41	4.76	5.02	5.24	5.43	5.59	5.74
10	3.15	3.88	4.33	4.65	4.91	5.12	5.30	5.46	5.60
11	3.11	3.82	4.26	4.57	4.82	5.03	5.20	5.35	5.49
12	3.08	3.77	4.20	4.51	4.75	4.95	5.12	5.27	5.39
13	3.06	3.73	4.15	4.45	4.69	4.88	5.05	5.19	5.32
14	3.03	3.70	4.11	4.41	4.64	4.83	4.99	5.13	5.25
15	3.01	3.67	4.08	4.37	4.59	4.78	4.94	5.08	5.20
16	3.00	3.65	4.05	4.33	4.56	4.74	4.90	5.03	5.15
17	2.98	3.63	4.02	4.30	4.52	4.70	4.86	4.99	5.11
18	2.97	3.61	4.00	4.28	4.49	4.67	4.82	4.96	5.07
19	2.96	3.59	3.98	4.25	4.47	4.65	4.79	4.92	5.04
20	2.95	3.58	3.96	4.23	4.45	4.62	4.77	4.90	5.01
24	2.92	3.53	3.90	4.17	4.37	4.54	4.68	4.81	4.92
30	2.89	3.49	3.85	4.10	4.30	4.46	4.60	4.72	4.82
40	2.86	3.44	3.79	4.04	4.23	4.39	4.52	4.63	4.73
60	2.83	3.40	3.74	3.98	4.16	4.31	4.44	4.55	4.65
120	2.80	3.36	3.68	3.92	4.10	4.24	4.36	4.47	4.56
infinity	2.77	3.31	3.63	3.86	4.03	4.17	4.29	4.39	4.47

Q Tables   (use for TUKEY Test)

Note: These are abbreviated tables with the most commonly used values. A more comprehensive table can be found in *pdfepXJ7Z5yxl*.

1st Page: Alpha = .05

2nd page: Alpha = .01

Q critical values for alpha = .05 df   for the Error Term

k= Number of Treatments

Q critical values for alpha = .01 df   for the Error Term

k = Number of Treatments

df↓ **k** →	2	3	4	5	6	7	8	9	10
5	5.70	6.98	7.80	8.42	8.91	9.32	9.67	9.97	10.24
6	5.24	6.33	7.03	7.56	7.97	8.32	8.61	8.87	9.10
7	4.95	5.92	6.54	7.01	7.37	7.68	7.94	8.17	8.37
8	4.75	5.64	6.20	6.62	6.96	7.24	7.47	7.68	7.86
9	4.60	5.43	5.96	6.35	6.66	6.91	7.13	7.33	7.49
10	4.48	5.27	5.77	6.14	6.43	6.67	6.87	7.05	7.21
11	4.39	5.15	5.62	5.97	6.25	6.48	6.67	6.84	6.99
12	4.32	5.05	5.50	5.84	6.10	6.32	6.51	6.67	6.81
13	4.26	4.96	5.40	5.73	5.98	6.19	6.37	6.53	6.67
14	4.21	4.89	5.32	5.63	5.88	6.08	6.26	6.41	6.54
15	4.17	4.84	5.25	5.56	5.80	5.99	6.16	6.31	6.44
16	4.13	4.79	5.19	5.49	5.72	5.92	6.08	6.22	6.35
17	4.10	4.74	5.14	5.43	5.66	5.85	6.01	6.15	6.27
18	4.07	4.70	5.09	5.38	5.60	5.79	5.94	6.08	6.20
19	4.05	4.67	5.05	5.33	5.55	5.73	5.89	6.02	6.14
20	4.02	4.64	5.02	5.29	5.51	5.69	5.84	5.97	6.09
24	3.96	4.55	4.91	5.17	5.37	5.54	5.69	5.81	5.92
30	3.89	4.45	4.80	5.05	5.24	5.40	5.54	5.65	5.76
40	3.82	4.37	4.70	4.93	5.11	5.26	5.39	5.50	5.60
60	3.76	4.28	4.59	4.82	4.99	5.13	5.25	5.36	5.45
120	3.70	4.20	4.50	4.71	4.87	5.01	5.12	5.21	5.30
infinity	3.64	4.12	4.40	4.60	4.76	4.88	4.99	5.08	5.16

END of TABLE Q

Chapter Eight

Simple Linear Regression

Regression Analysis refers to mathematical models in which variables are statistically related, i.e., the value of one variable can be estimated on the basis of the values of other variables.

Distinction Between ANOVA and Regression

The difference between ANOVA and Regression is illustrated in the following example:

We want to investigate the effect of the number of patrol cars assigned to a neighborhood on the number of reported burglaries. In the study, different numbers of patrol cars are assigned to similar neighborhoods in a large city. The <u>dependent variable</u> is the number of burglaries reported in each neighborhood during the study period. The <u>independent variable</u> is the number of patrol cars assigned to the neighborhood. Suppose we used three levels of the independent variable - 2, 4, and 6 cars. Remember that we assume a probability distribution for the number of reported burglaries at each level of the independent variable. We assume that the distributions are normal with equal variances. Only the means may differ. Suppose our ANOVA model looks as follows:

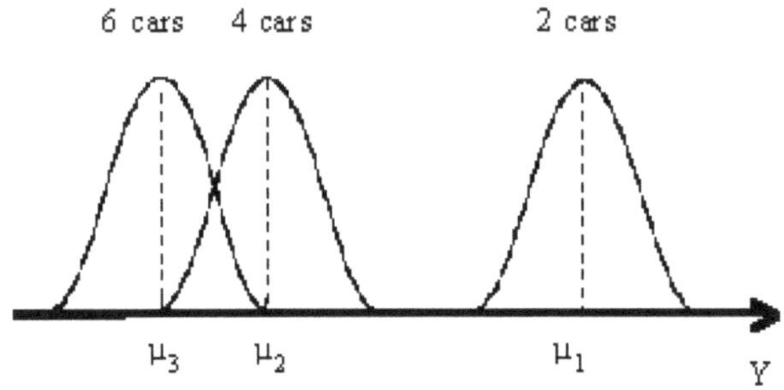

From the picture, we know:

(1) The mean number of burglaries with two cars appears to be much higher than for 4 or 6 cars.

(2) The decline in the mean number of burglaries when the number of cars is increased from 4 to 6 is smaller than from 2 to 4.

Now, let's look at this example problem as a Regression Analysis (ANOR):

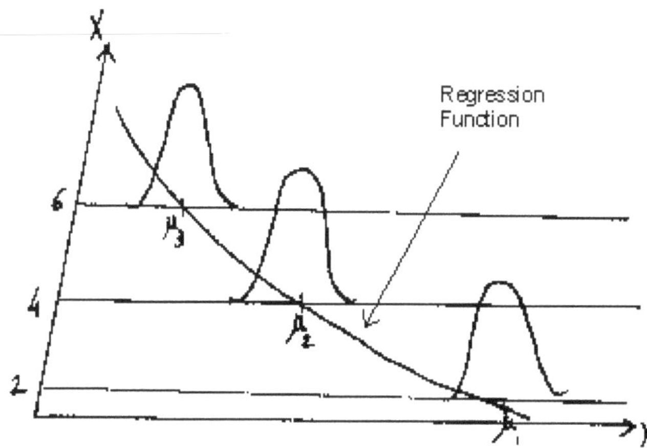

The <u>Regression Function</u> provides information about the expected number of burglaries for intermediate levels of patrol cars (3 and 5 cars) even though we didn't test at these levels because it assumes a <u>Response Curve</u>. (ANOVA provides no such intermediate information. ANOVA only provides information about the levels studied - 2, 4, and 6 cars.) The <u>dependent</u> and <u>independent</u> variables in a Regression model are usually <u>quantitative.</u> In an ANOVA, the <u>dependent</u> variable is <u>quantitative,</u> but the <u>independent</u> variable is usually <u>qualitative</u>. In the soft drink marketing example, the independent variable is the <u>color</u> of the drink (qualitative). Thus the soft drink problem is an ANOVA, <u>not</u> a Regression. For example, what intermediate level is there between pink and green? Green and orange?

Simple Linear Regression Model

(Dependent Variable) (Independent Variable)

$$\Downarrow \qquad\qquad \Downarrow$$

$$Y_i \;=\; \beta_0 \;+\; \beta_1 X_{1i} \;+\; \epsilon_i$$

$$\uparrow \quad \uparrow \qquad \uparrow$$

(Y-intercept) (Slope)   (Error Term)

The model is a <u>population</u> (notice the Greek β's and $\in$). It is a "***simple***" (meaning <u>one</u> X variable) and a "***linear***" (meaning straight line in β's) regression model. (This is a fancy way to write the equation for a straight line, Y = A + BX .)

In Regression, Y is dependent on the value that X_1 assumes. (This is the reason that X_1 is called the <u>independent</u> variable and Y is the <u>dependent</u> variable.) β_0 and β_1 will be numbers. β_1 can be +, - , or 0.

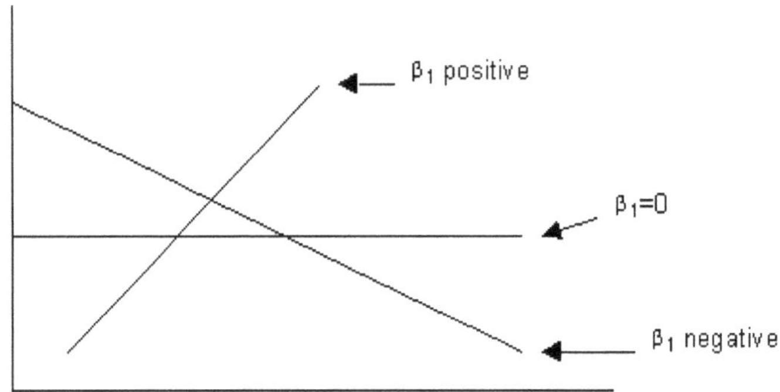

Remember that the model assumes a linear relationship between Y and X_1. If we take the <u>expected value</u> (average) of the Regression model:

$$E(Y_i) = \beta_0 + \beta_1 X_1$$

<u>Notes:</u>

(1) The $\in_i$ disappears because E ($\in_i$) = 0; that is, we expect our error to be 0 **(2)** This model indicates that given a level of X_1, you can generate the mean of the probability distribution of the Y's at that X_1 level.

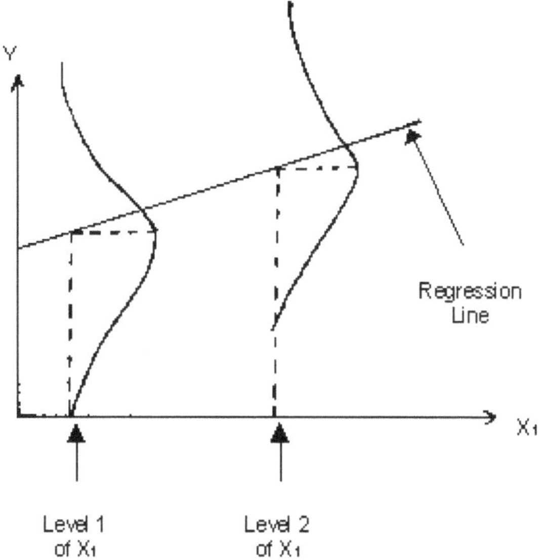

<u>Note:</u> Normal distributions with <u>equal variance</u> occur at each level of X_1.

What are the ϵ_i's? Statistical relationships are not perfect, so they represent the deviation of the individual Y_i's from the Regression line.

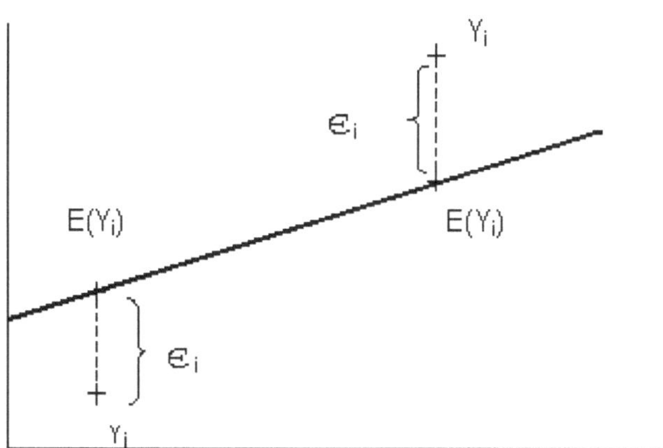

Some ϵ_i's are negative, some positive, and some zero. Remember, $E(\epsilon_i)=0$, which, is the average of the errors is zero. In a nutshell, the regression line represents the average, and the error term represents how far the individual observation is from the average.

Sample Regression Line

We must estimate the population Regression line using <u>sample</u> data. The <u>sample regression line</u> is:

$$\hat{Y}_i = b_0 + b_1 X_{1i}$$

Where $e_i = Y_i - \hat{Y}_i$

(actual value)   (predicted value)

Note:

1) No Greek letters.

2) The ^ on the Y_i means that it has been estimated from sample data.

Least Squares Method

In order to "fit" a regression line to sample data, you must find the line that minimizes $\Sigma\, e_i^2$. (Hence the name least squares.)

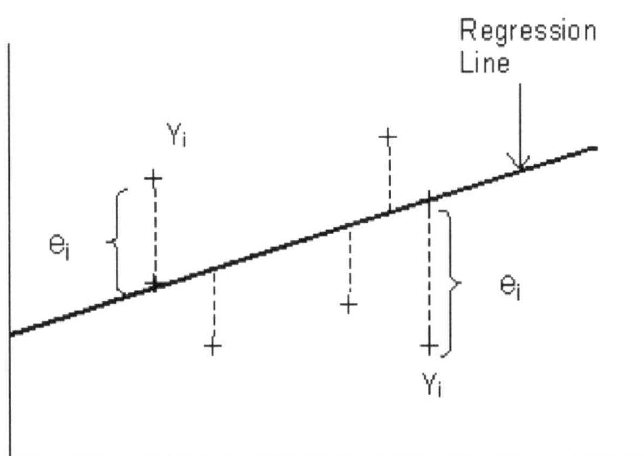

Remember, some e_i's are positive, some negative, and some 0. $\Sigma\, e_i = 0$. The e_i's must be squared in order to do away with the negative signs and ensure that you get the "line of best fit." Using calculus the equations for b_0 and b_1 which min $\Sigma\, e_i^2$:

$$b_1 = \frac{n\sum X_1 Y - \sum X_1 \sum Y}{n\sum X_1^2 - \left(\sum X_1\right)^2}$$

$$b_0 = \frac{\sum Y}{n} - b_1 \frac{\sum X_1}{n}$$

$$= \overline{Y} - b_1 \overline{X_1}$$

Example

A teacher feels that hours/week doing homework, X_1, is a useful predictor of exam scores, Y. (Given data for X_1:   5   10   4   0   6

As follows (respectively):

	Data			Calculations
i	X	Y	X_1^2	$X_1 Y$
1	5	70	25	350
2	1	80	100	800
3	4	70	16	280
4	0	60	0	0
5	6	70	36	420

(a) Estimate the regression line.

$\sum X_1 = 25$ $\sum X_1 Y = 1850$ $\sum Y = 350$ $\sum X_1^2 = 177$ $n = 5$

$$b_1 = \frac{5\,(1850) - 25\,(350)}{5\,(177) - (25)^2} = 1.92$$

$$b_0 = \frac{350}{5} - 1.92\left(\frac{25}{5}\right) = 60.4$$

$$\hat{Y}_i = 60.4 + 1.92\,X_{1i}$$

(b) If a student <u>increases</u> hours/wk doing homework by 2 hours/wk, by how much would we <u>expect</u> the student's exam grade to change? (Give magnitude & direction)

$b_1 X_1 = (1.92)\,(2) = +3.84\,.\,+4$ pts.

(c) Interpret b_0.

This is the expected exam score given <u>no</u> (zero) hrs/wk doing homework.

(d) Predict the expected exam score for a student who spends 8 hours/week doing homework.

$$\hat{Y} = 60.4 + (1.92)\,8 = 75.76 \approx 76$$

(e) Would you expect this relationship to be linear over all possible values of X_1 and Y?

No. You would expect to reach a point of diminishing returns where more hours/week doing homework would not produce as much improvement in exam scores.

(f) After calculating $\hat{Y}$ (using estimate equation $\hat{Y} = \mathbf{60.4 + 1.92\ X_1}$ Calculate the e_i 's (the discrepancies between Y and $\hat{Y}$. What do they represent?

i	Y_i	$\hat{Y}_i$	e_i
1	70	70	0
2	80	79.6	.4
3	70	68.08	1.92
4	60	60.4	-.4
5	70	71.92	-1.92

$$\sum e_i = 0$$

Note: **(1)** Y_i are the actual data values.

(2) $\hat{Y}_i$ are the predicted values using the regression model - input the X_1

Data into the regression model and output the predicted $\hat{Y}$'s.

(3) $\quad e_i = Y_i - \hat{Y}_i$

The error term, e_i, represents how far off your prediction is for each data point used in developing the model. Notice that the error term has a magnitude and a direction. For example, for student #5, the regression model predicted an exam score of 71.92. The student actually scored 70. The prediction was too high by 1.92 points, i.e., e_5=-1.92.

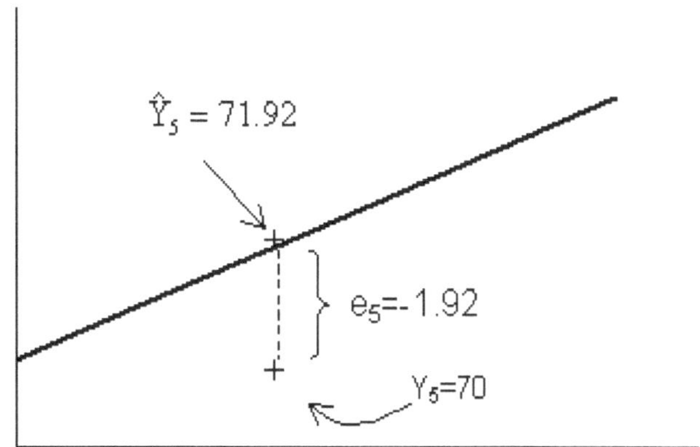

(g) Calculate R^2, the Coefficient of Determination. Interpret R^2.

$$R^2 = \left(\frac{n\sum X_1 Y - \sum X_1 \sum Y}{\sqrt{[n\sum X_1^2 - (\sum X_1)^2][n\sum Y^2 - (\sum Y)^2]}} \right)^2$$

R^2 indicates the strength of the relationship between X_1 and Y. ($0 \leq R^2 \leq 1$)

R^2 is the proportion of variation in Y accounted for by X_1. The closer R^2 is to 1, the higher the degree of statistical relation between Y and X_1.

$$\sum Y^2 = 24{,}700$$

$$R^2 = \left(\frac{(5)\,(1850) - (25)\,(350)}{\sqrt{[(5)\,(177) - (25)^2][(5)\,(24700) - (350)^2]}} \right)^2$$

$$= (.98)^2 = .96$$

96% of the variation in exam score (Y, the dependent variable) is accounted for by hours/week doing homework (X_1, the independent variable).

<u>Note:</u> $\sqrt{R^2} = r$ which is the correlation coefficient. r has the same sign as b_1. ($-1 \leq r \leq 1$)

Partitioning the Sum of Squares and Degrees of Freedom

Just as in ANOVA, the SS, and d.f. can be partitioned in Regression.

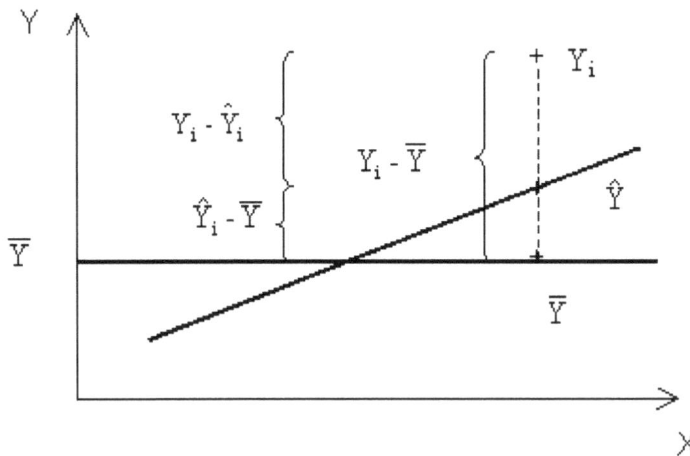

Now square the components and sum across all data points:

$$\sum (Y_i - \overline{Y})^2 = \sum (\hat{Y}_i - \overline{Y})^2 + \sum (Y_i - \hat{Y}_i)^2$$

$$SSTO = SSR + SSE$$

Note: SSR = Sum of Squares Regression

To Partition the d.f.:

(1) SSTO has n - 1 d.f. because $\sum(Y_i - \overline{Y}) = 0$ is a constraint.

(2) SSR has p - 1 d.f. where p is the number of β's in the model. We lose a d.f. because of the constraint $\sum(\hat{Y}_i - \overline{Y}) = 0$.

(3) SSE has n - p d.f because there are p constraints, and one is associated with each $\square$ estimate.

Summary:

$$(n - 1) \quad = \quad (p - 1) \quad + \quad (n - p)$$

d.f. for SSTO   d.f. for SSR   d.f. for SSE

To calculate the Mean Squares:

MSR = SSR / p -1

MSE = SSE / n − p

To calculate F*:

$F^*_{p-1, n-p}$ = MSR / MSE

ANOR Table

Source	d.f.	SS	MS	F
Regression	p - 1	SSR	MSR	F*
Error	n - p	SSE	MSE	
Total	n - 1	SSTO		

Note: R^2 = SSR/SSTO

The F Test

In simple regression, the F tests the hypothesis:

H_0: $\beta_1 = 0$

H_A: $\beta_1 \neq 0$

If $\beta_1 = 0$, there is <u>no</u> relationship between Y and X_1. This means that the regression model is not useful.

<u>**CI for β_1**</u>

The CI for β_1 is:

$$b_1 - t_{\alpha/2,n-p} \sqrt{\frac{MSE}{\sum (X_i - \overline{X})^2}} \leq \beta_1 \leq b_1 + t_{\alpha/2,n-p} \sqrt{\frac{MSE}{\sum (X_i - \overline{X})^2}}$$

Example Problem ANOR

Source	d.f.	SS	MS	F
Regression	1	192	192	71.91
Error	3	8	2.67	
Total	4	200		

$\overline{Y} = 350 / 5 = 70$

$SSR = \sum (\hat{Y}_i - \overline{Y})^2$

$= (70 - 70)^2 + (79.6 - 70)^2 + (68.08 - 70)^2 + (60.4 - 70)^2 + (71.92 - 70)^2$

$= 191.6928. \ 192$

$SSTO = \sum (Y_i - \overline{Y})^2$

$= (70 - 70)^2 + (80 - 70)^2 + (70 - 70)^2 + (60 - 70)^2 + (70 - 70)^2$

$= 200$

SSE can be obtained by subtraction, i.e., SSTO - SSR = SSE.

$SSE = 7.6928. \ 8$

$n = 5 \quad p = 2 \quad n - 1 = 4 \quad p - 1 = 1 \quad n - p = 3$

$MSR = 192 / 1 = 192$

$MSE = 8 / 3 = 2.67$

F* = 192 / 2.67 = 71.91

$H_0: \beta_1 = 0$

$H_A: \beta_1 \neq 0$

$F_{.05, 1, 3} = 10.1 < F^*$

Reject $H_0 \rightarrow \beta_{ten}$ and X_1 can be used to predict Y.

Calculate a 95% CI for β_1: $\quad 1.92 \pm t_{.025,3} \sqrt{\dfrac{2.67}{52}} = 1.92 \pm 3.182\,(.2266) = 2.641, 1.199$

$$P\,(1.199 \leq \hat{\beta_1} \leq 2.641) = .95$$

Note:	X_i	$(X_i \quad - \quad \overline{X})^2$
	5	0
	10	25
	4	1
	0	25
	6	1
$\overline{X} =$	5	$\Sigma\ (X_i - \overline{X})^2 = 52$

<u>NOTE</u>

Both endpoints of the CI are positive $\beta \neq 0$ same results as the hypothesis test.

Time Series Simple Linear Regression Models

I.   Trend Model

$Yt = \beta_0 + \beta_1 Xt$

$\uparrow$

    (Linear time trend)

where Xt is a coded time variable $X_1 = 1$, $X_2 = 2$, etc.

II.   Lagged Model

Ex/ $Y_t = \beta_0 + \beta_1 X_{t-5}$

If Y_t is kindergarten enrollment in year t and X_{t-5} is the number of births five years earlier, then X_{t-5} is a lagged time series relative to Y_t. The lag is five years.

<u>Note:</u> In time series, the formulas for calculating b_0 and b_1 are the same as used previously.

Note on SAS Input

Suppose you want to make predictions with your regression model and input only the independent measure for that observation with <u>no</u> corresponding dependent measure. SAS uses the observation for prediction purposes <u>only</u> (<u>not</u> for model development).

Ex/   CARDS;

                            570

                           1080

                            470

                            060

                            670

8   ← SAS calculates predicted exam scores for students with 8 hrs/wk homework

TITLE ' ... ';

Notes on Regression Printouts

Example

(1) $R^2 = .961538$ implies 96% of the variation in the score is attributed to hours.

(2) Regression Model : $\hat{Y} = 60.38 + 1.92 X_1$

(3) PR > |T| for slope = PR > F for model because both test $H_0: \beta_1 = 0$ $H_A: \beta_1 \neq 0$

(4) (T for H_0)^2 = F Value. Remember, the t for the CI has d.f. for MSE, and the F has d.f. MSR, MSE. Ex/ $(t_3)^2 = F_{1,3}$

(5) Standard Error of Estimate =

Therefore a CI for β_1 $1.92 \pm t_{\square/2,\,3}\,(.22)$

$$\uparrow$$

(Looked up in t $\sqrt{\dfrac{MSE}{\sum (X_i - \bar{X})^2}}$ table)

(6) Observed value = Y_i

Predicted Value = $\hat{Y}_i$

Residual = $(Y_i - \hat{Y}_i)$

Notice: Sum of Residuals = 0

(7) Always compare Root MSE = $\sqrt{MSE}$ with Y variable mean. This gives you an idea of how accurate your model is. Here Root MSE = 1.60 vs. score mean = 70.0, and Root MSE is <u>much</u> smaller than the mean. This means that the normal distributions sitting over the regression line are tall and narrow as opposed to short and wide. (You don't want short and wide because the variance is too big, and your predictions will be subject to large errors.)

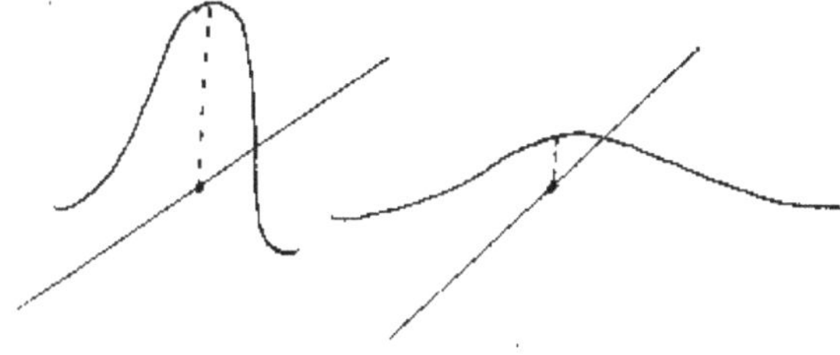

(8) When the Residuals are plotted against X1, you should see a random scatter about the "zero" line:

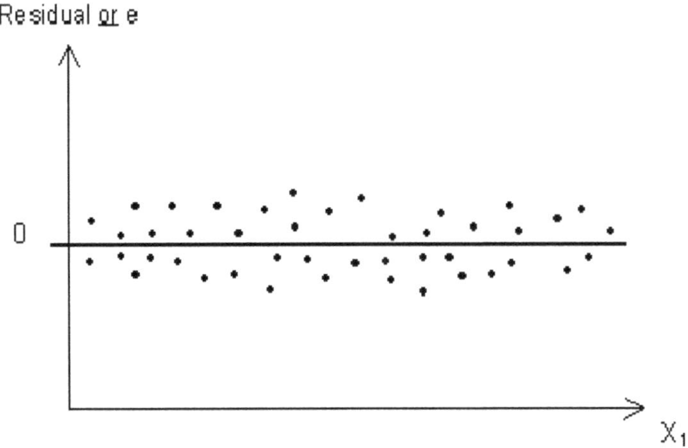

If the pattern is not a random scatter, this requires revision to the original model.

Ex/

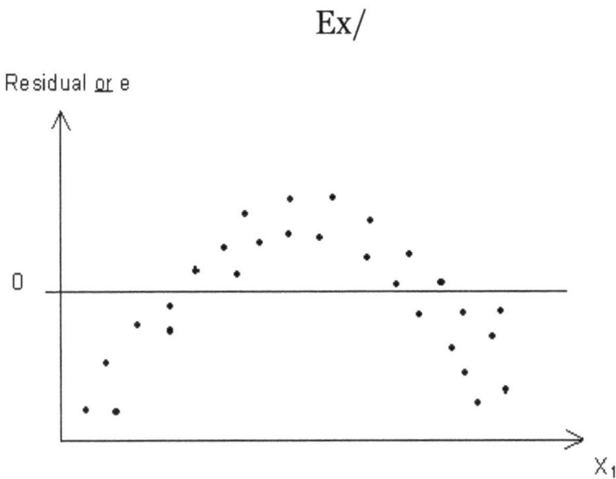

This pattern indicates that a quadratic model is more appropriate than a linear model.

$$Y = \beta_0 + \beta_1 X_1 + \beta_2 X_1^2$$

E X C E L Computations

The EXAMPLE of Teacher's scores   and hours spent on Homework

EXCEL   parameters (coefficients) estimates

<table>
<tr><td>

$$b_1 = \frac{5\,(1850) - 25\,(350)}{5\,(177) - (25)^2} = 1.92$$

$$b_0 = \frac{350}{5} - 1.92\left(\frac{25}{5}\right) = 60.4$$

$$\hat{Y}_i = 60.4 + 1.92\,X_{1i}$$

</td>
<td style="border:1px solid black; padding:1em;">

if not rounded

1.92

60.38

-

$Y{\char`\^}i = 60.385 + 1.9231\,X_{1i}$

</td></tr>
</table>

$\underline{X_1}$	$\underline{Y}$	$Y{\char`\^}$	Y- Ybar	Y^- Ybar	Y-Y^	(Y- Ybar) squared	(Y^- Ybar) Squared	(Y-Y^) squared
5	70	70	0	0	0	0	0	0

10	80	79.61	10	9.61	0.38	100	92.45	0.14
4	70	68.07	0	-1.92	1.92	0	3.69	3.69
0	60	60.38	-10	-9.61	-0.38	100	92.45	0.14
6	70	71.92	0	1.92	-1.92	0	3.69	3.69
Sum s: 25	350	350	0	0	0	200	192.3	7.69
						SST	**SSR**	**SSE**

As you can see :

$$SST = SSR + SSE$$

$$200 \quad 192.31 \quad 7.69$$

ANOVA Table

Variation Source	Sum of Squares	degrees of freedom	Mean Square	F* Calculated	p-value
Regression	192.3		192.3	75.00	
Error	7.60		2.56		
TOTAL	200	4			

Chapter Nine

Multiple Linear Regression Modelling

Introduction

What the mastery of linear models entails will depend on your field and the specific type of research you do. The focus of this section is on statistical modeling, which basically having the same core structure. There is always a response variable, a set of predictors, an estimate of the nature of their relationship, and a residual term (responsible for the "randomness" or lack of it). The details of each of the models will vary, but if you can truly understand one basic type, any other type of modeling is a few steps away.

The previous chapters took you up through simple linear regression and ANOVA; the following part is about mastering deeper aspects of higher-order linear modeling. That will include extra concepts of <u>dummy   variables</u>, <u>interactions</u>, <u>polynomial   effects</u>, <u>random   effects</u>, model building, <u>model fit</u>, etc. To truly master this stage means a thorough understanding of <u>how ANOVA and regression fit together</u> into the General Linear model and being able to fluently move from one to the other. The parts that are not covered in this chapter are structural equation modeling, multivariate statistical methods, <u>survival analysis</u>, and <u>complex survey</u> techniques, among others.

Software is important and is essential for researchers to know, and the methods used in the software will, of course, be more sophisticated, and any researcher should have a real understanding of the program's defaults, vocabulary, and how each bit of output is interpreted. It is a great idea to be thoroughly knowledgeable in one package/software you need it. Many researchers get more savvy in the software when they hit the dissertation stage. Unfortunately, such a strategy gets researchers in trouble a lot. They simply lack the experience, and that is never amenable through textbook examples, a researcher needs to dirty his/her hands with real software practices with different data sets and models with different research questions and hypotheses, and it may take years to gain experience with a variety of models.

Multiple Regression

Multiple regression is an extension of simple regression where more than one independent (X) variable is involved.

First Order, Two Independent Variables

$Y_i = \beta_0 + \beta_1 X_{i1} + \beta_2 X_{i2} + \varepsilon_i$ **(population)** <u>OR</u> the estimate:

$$\hat{Y}_i = b_0 + b_1 X_{i1} + b_2 X_{i2} \qquad \text{(sample)}$$

This model yields a <u>response surface</u> that is a plane (rather than a straight line as in simple linear regression.)

β_0 = Y intercept of the regression plane $(X_1 = 0, X_2 = 0)$

β_1 = Change in mean response (Y) per unit change in X_1 <u>when X_2 is held constant</u>

β_2 = Similar interpretation to β_1

β_1 and β_2 are partial regression coefficients. They represent the partial effect of one independent variable when the other is held constant.

First Order, P-1 Independent Variables

$Y_i = \beta_0 + \beta_1 X_{i1} + \beta_2 X_{i2} + \ldots + \beta_{p-1} X_{p-1} + \varepsilon_i$ (population) <u>OR</u>

$$\hat{Y}_i = b_0 + b_1 X_{i1} + b_2 X_{i2} + \ldots + b_{p-1} X_{p-1} \qquad \text{(sample)}$$

This response surface is a <u>hyperplane</u> (more than two dimensions). The $\square$'s are interpreted as before, where $\square_k$ is the change in mean response per unit change in X_k when all the other independent variables are held constant.

General Linear Models (GLM)

Most regression models fall into the GLM category. GLMs include:

(1) Additive models (no interaction terms). All of the simple and multiple regression models that we have discussed so far are of this type.

(2) Polynomial models (some of the X variables are raised to a power)

Ex/ $Y_i = \beta_0 + \beta_1 X_1 + \beta_2 X_1^2 + \varepsilon$

(3) Transformed models (Here, you may take logs or reciprocals, etc., of some of the X variables or the Y variable).

(4) Interaction models (here, you have X variables multiplied by each other).

Ex

$Y_i = \beta_0 + \beta_1 X_1 + \beta_2 X_2 + \beta_3 (X_1 X_2) + \varepsilon$

Note: For this type of model, the change in the mean response when X_1 changes by one unit (X_2 held constant) are $\beta_1 + \beta_3 X_2$ (and similarly for X_2).

(5) More complex models (any combination of the first four types)

Note: GLM encompasses both linear and nonlinear models. Ex/ Types (2) & (4) are nonlinear models

Least Squares

The $\square$'s in multiple regression are estimated using the least squares method. Since we are working with more than one X variable, we must use matrix algebra to solve for b_0, b_1, b_2, etc.

ANOR Table

Looks just like simple regression:

Source	d.f.	SS	MS	F
Regression	p - 1	SSR	MSR	F*
Error	n - p	SSE	MSE	
Total	n - 1	SSTO		

(Remember: p = the number of $\square$'s in the model.)

F Test for Overall Regression Model

The hypothesis that we are testing in a multiple regression model is:

$H_0: \beta_1 = \beta_2 = \ldots = \beta_{p-1} = 0$

H_A: Not all β 's = 0

(Note: This is not the same as $\beta_1 \neq \beta_2 \neq \ldots \neq \beta_{p-1}$. Don't write H_A this way!)

As always, we are looking for an F* that is much > more than 1.

Coefficient of Multiple Determination

$R^2 = SSR/SSTO$

Again $0 \leq R^2 \leq 1$.

Remember: Even with a high R^2, if MSE is large, your model will not be very precise.

CI for Individual β 'S

If H_A is concluded, you know that <u>at least one β is not zero</u>. But which ones or ones are not zero?

$$b_k - t_{\alpha/2, n-p} \sqrt{\frac{MSE}{\sum (X_{ik} - \bar{X}_k)^2}} \leq \beta_k^2 \leq b_k + t_{\alpha/2, n-p} \sqrt{\frac{MSE}{\sum (X_{ik} - \bar{X}_k)^2}}$$

We can use CIs to find out the following:

With the CIs we are testing:

H_0: $\beta_k = 0$

H_A: $\beta_k \neq 0$

Using the Multiple Regression Model for Predictions

Just as in simple regression, a multiple regression model may not be appropriate outside the range of the observed values. In addition, in multiple regression, even if a data point is within the range of each X variable, this does not mean that the point is on the regression response surface.

Ex/

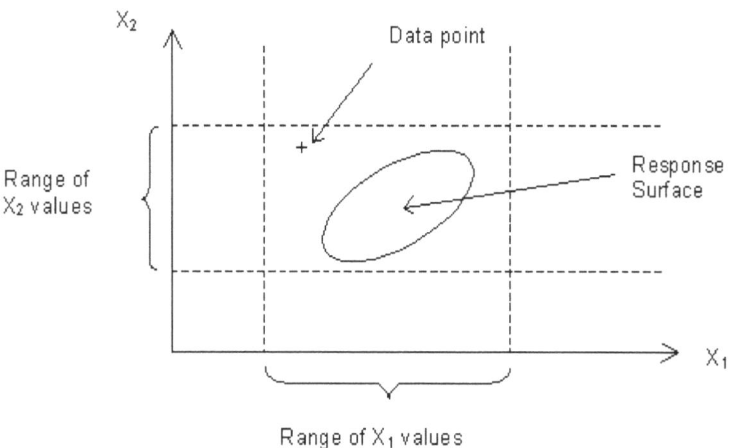

Multicollinearity

One objective of regression models may be to measure the <u>separate effects</u> of the <u>independent</u> variables on the <u>dependent</u> variable. Generally, the β indicates the change in the mean of Y when X changes by one unit and the other independent variables are held constant. However, if the sample observations of the <u>independent</u> variables are <u>highly correlated</u> among themselves, the separate effects cannot be measured by the β's. This problem is known as <u>multicollinearity</u>.

Multicollinearity does not adversely affect the regression equation if the purpose of your research is only to predict the dependent variable from a set of predictor variables. In this case, the predictions in your regression will still be accurate, and the overall $R2$ will give you an indication of how well the predictor variables in your model predict the dependent variable. Multicollinearity does not affect the goodness of fit and the goodness of prediction. In regression, multicollinearity can be a problem if the purpose of your study is to estimate the contributions of individual predictors. When multicollinearity is present, p values can be misleading, and the regression coefficients' confidence intervals will be very wide and may vary dramatically with the addition or exclusion of just one case/participant. If this is the case, removing any highly correlated terms from the model will greatly affect the estimated coefficients of the other highly correlated terms. Multicollinearity inflates the variances of the parameter (β_0, β_1, ... , β_p) estimates. This may lead to a lack of statistical significance of individual independent variables even though the overall model may be significant. This is especially true for small and moderate sample sizes. Such problems will result in incorrect conclusions about relationships between independent and dependent variables. This is a mistake you don't want to make in your study.

Multicollinearity in regression occurs when predictor variables (independent variables) in the regression model are more highly correlated with other predictor variables than with the dependent variable. Multicollinearity does not adversely affect the regression equation if the purpose of your research is only to predict the dependent variable from a set of predictor variables. In this case, the predictions in your regression will still be accurate, and the overall R2 will give you an indication of how well the predictor variables in your model predict the dependent variable. Multicollinearity does not affect the goodness of fitness and the goodness of prediction.

In regression, multicollinearity can be a problem if the purpose of your study is to estimate the contributions of individual predictors. When multicollinearity is present, p values can be misleading, and the regression coefficients' confidence intervals will be very wide and may vary dramatically with the addition or exclusion of just one case/participant. If this is the case, removing any highly correlated terms from the model will greatly affect the estimated coefficients of the other highly correlated terms. Multicollinearity inflates the variances of the parameter (β, 1, , β_p) estimates. This may lead to a lack of statistical significance of individual independent variables even though the overall model may be significant. This is especially true for small and moderate sample sizes. Such problems will result in incorrect conclusions about relationships between independent and dependent variables. This is a mistake you don't want to make in your study.

(<u>Note</u>: Correlations are easily obtained by adding a PROC CORR; statement to your SAS jobs before the PROC GLM;)

Example: Suppose we have two variables, X_1, and X_2, which are highly correlated (r_{12} = .99).

(<u>Note</u>: Correlation is denoted by r and $-1 \leq r \leq +1$. The closer r is to -1 or +1, the stronger the degree of the linear relation between the two variables.)

Run all three models (simple and multiple regression):

Slope	$\hat{Y} = b_0 + b_1 X_1$	$\hat{Y} = b_0 + b_2 X_2$	$\hat{Y} = b_0 + b_1 X_1 + b_2 X_2$
b_1	34.86	---	57.75
b_2	---	.35	-.23

Notice how the b's change when both variables are included in the model. Multicollinearity also affects the t-tests on the individual $\square$'s.

$H_0: \beta_k = 0.$ $\rightarrow$ Remember, we used CIs to test this

$H_A: \beta_k \neq 0.$

When multicollinearity exists, the F test leads you to H_A, but all of the CIs indicate H_0. This is inconsistent since at least one $\square$ has to be nonzero. Why does this happen? The t-test or CI measures the marginal contribution of an X variable, given that all other X variables are already included in the model. Say X_1 and X_2 are highly correlated, then the marginal contribution of X_2 to the reduction in variation of Y when X_1 is included in the model is minimal. Thus $\beta_2 = 0$. And vice versa when you look at β_1!

How do we remedy the situation? Not easily. You can try to eliminate some of the X variables or combine them into surrogate variables using advanced statistical techniques.

Indicator or Dummy Variables

The X variables in multiple regression are sometimes qualitative.

Ex/ $X_1 = 0$ if male <u>or</u>

        $= 1$ if female

X_1 is called an indicator variable, and β one measures the effect of sex on E(Y).

Residual Analysis

The ei's or residuals should have a random scatter about the "zero" line. When using multiple regression, the I' must be plotted against each X variable.

<u>**EX 1/**</u>	X_1 Media Expenditures	X_2 Point of Sale Expenditures	Y Sales Volume
<u>i</u>	<u>($1000)</u>	<u>($1000)</u>	<u>($10,000)</u>
1	2	2	8.74
2	2	3	10.53
3	2	4	10.99

.	.	.	.
.	.	.	.
.	.	.	.
16	5	5	20.51

CORRELATION MATRIX

	X_1	X_2	Y
X_1	1.000	.000	.965
X_2		1.000	.225
Y			1.000

No problem with multicollinearity here. X_1 is more closely associated with Y than X_2.

Parameter	Estimate	T
Intercept	2.13438	3.50
X1	3.02925	25.18
X2	.70575	5.87

Table t value = t_{13} = 2.160

Both β_1 and β_2 $\neq$ 0.

ANOR

Source	d.f.	SS	MS	F
Model	2	193.4888	96.74439	334.35
Error	13	3.7616	0.28935	
Total	15	197.2503		

Table F value = $F_{2,13}$ = 3.81

(Note: The F test is for H_0: $\beta_1 = \beta_2 = 0$

H_A: Not all β's = 0

The t tests are $H_0: \beta_1 = 0$ $H_0: \beta_2 = 0$

             $H_A: \beta_1 \neq 0$ $H_A: \beta_2 \neq 0$

The regression model is $\hat{Y} = 2.13438 + 3.02925\,X_1 + .70575\,X_2)$

$R^2 = .98093$ $\rightarrow$ Looks like a good fit

Root MSE = .53791 $\nearrow$

Residuals Analysis

	Y_i	$\hat{Y}_i$	e_i
i	Observed	Predicted	Residual
1	8.74	9.60	-.86
2	10.53	10.31	.22
3	10.99	11.02	-.03
.	.	.	.

(We would need to plot these to look for patterns)

<u>Interpretation of b_0, b_1, b_2:</u>

b_0 = Sales volume of 2.13438 if no advertising of any kind ($X_1 = 0$, $X_2 = 0$)

B_1 = If Media Expenditures increase by 1 unit and <u>POS Expenditures remain constant</u>, sales volume will increase by 3.02925. You do b_2.

<u>**EXAMPLE 2/**</u>

	X_1	X_2	X_3	Y
	Manual Dexterity	Depth Perception	$(X_2)^2$	Job Proficiency
i	Score	Score		Score
1	60	82	6724	1102
2	135	163	26569	2333
3	101	61	3721	1384
.	.	.	.	.
.	.	.	.	.
.	.	.	.	.
25	77	122	14884	1513

CORRELATION MATRIX

	X_1	X_2	X_3	Y
X_1	1.000	.796	.781	.950
X_2		1.000	<u>.977</u>	.945
X_3			1.000	.924
Y				1.000

Potential problem because X_3 is really $(X_2)^2$, <u>highly related</u> to X_2.

Parameter	Estimate	T
Intercept	.08042	.02
X_1	10.00907	219.14
X_2	6.12105	72.96
X_3	-0.00045	<u>-1.36</u>

Table t value = t_{21} = 2.080 $\rightarrow$ $\square_3$ = 0

ANOR (or call it ANOVA)

Source	d.f.	SS	MS	F
Model	3	10,735,330	3,578,443	149,535
Error	21	502	24	
Total	24	10,735,832		

Table F value = $F_{3,21}$

R^2 = .99

Root MSE = 4.89

(<u>Note</u>: The F test is for H_0: $\beta_1 = \beta_2 = \beta_3 = 0$

$\qquad\qquad H_A$: Not all $\square$'s = 0

The t test are H_0: $\beta_1 = 0 \qquad H_0$: $\beta_2 = 0 \qquad H_0$: $\beta_3 = 0$

$\qquad\qquad H_A$: $\beta_1 \neq 0 \qquad H_A$: $\beta_2 \neq 0 \qquad H_A$: $\beta_3 \neq 0$

The regression model is $\hat{Y} = .08042 + 10.00907X_1 + 6.12105X_2 - .00045X_3$)

<u>Residuals</u> (You fill in)

i	Y_i Observed	$\hat{Y}_i$ Predicted	e_i Residual
1	_______	_______	_______
2	_______	_______	_______
3	_______	_______	_______
.	.	.	.
.	.	.	.
.	.	.	.

Since $\beta_3 = 0$, we would eliminate X_3 from our model and rerun the SAS job.

SSI and SSII

SSI - Sequential (Additive)

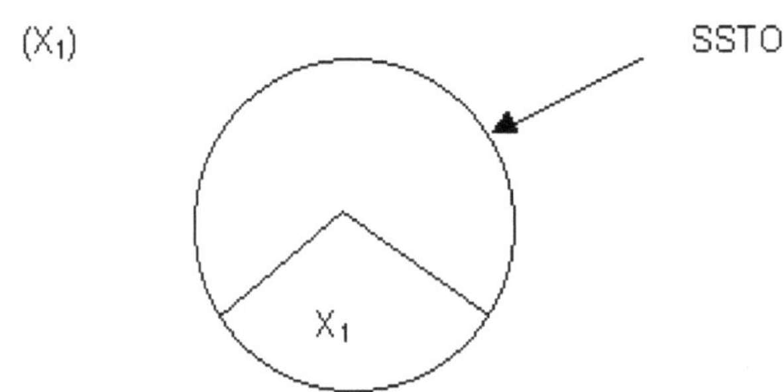

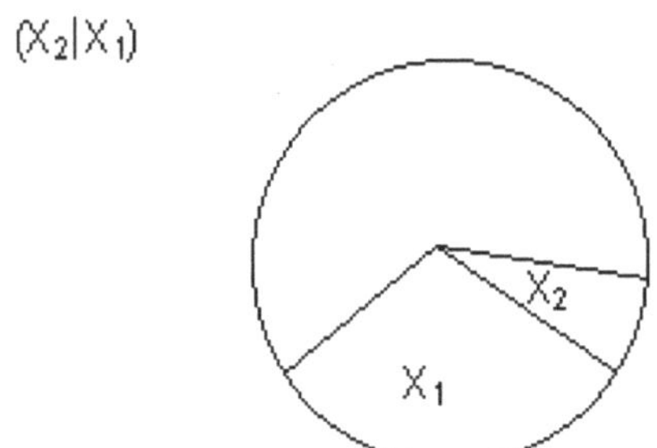

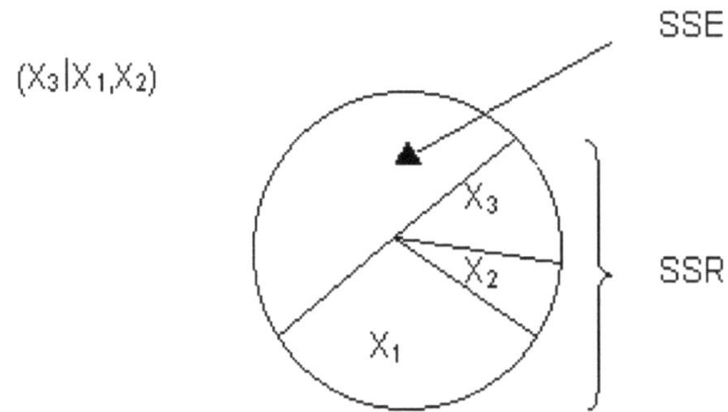

SSII - Marginal (Not additive) (t-tests)

$(X_1|X_2,X_3)$

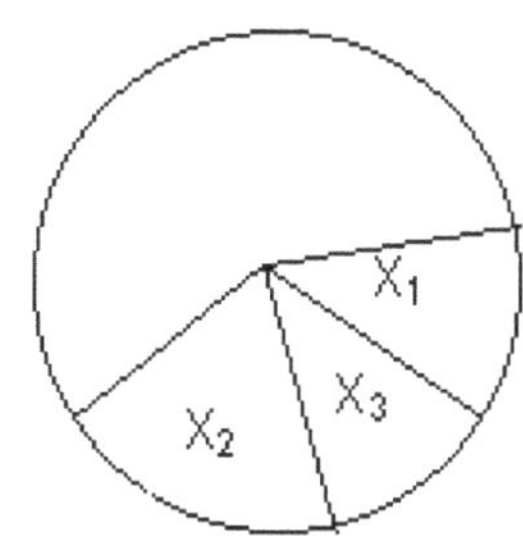

$(X_2|X_1,X_3)$

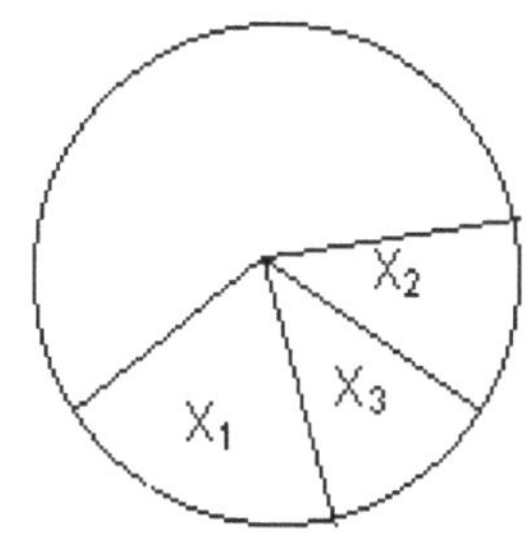

$(X_3|X_1,X_2)$

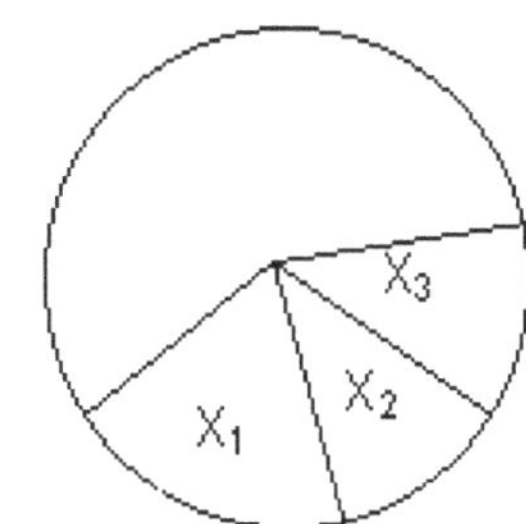

MULTIPLE REGRESSION--KNOW HOW TO:

(1) Interpret R2.

(2) Do CI's for the β's.

(3) Make predictions with the model.

(4) Write out the model.

(5) Know how R2 is calculated, SSR/SSTO.

(6) Know how to calculate residuals.

(7) Interpret the intercept and slopes.

Two-Way Analysis of Variance

<u>F distribution and F-test</u> - Two-Way Analysis of Variance

Two-Way ANOVA:

Two-way analysis of variance is the extension of one-way analysis of variance. There are two independent variables (therefore, the name is two-way).

Assumptions:

a) The populations from which the samples were obtained should be normally or about normally distributed.

b) The samples should be independent.

c) The variances of populations should be equivalent.

d) The groups should have similar sample sizes.

Hypotheses:

In a two-way ANOVA, there are three sets.

The null hypotheses for each and every set are shown below:

a) The population means of the first factor are equivalent. This is similar to the one-way ANOVA for the row factor.

b) The population means of the second factor are equivalent. This is similar to the one-way ANOVA for the column factor.

c) There is no interaction between the two factors. This is identical to executing a test for independence with the contingency tables.

Factors:

In a two-way ANOVA, the two independent variables are termed as factors. The idea is that there are two variables or factors which influence the dependent variable. Each and every factor will encompass two or more levels in it, and the degree of freedom for all factors is one less than the number of levels.

Treatment Groups:

Treatment Groups are made by making all possible combinations of the two factors. For illustration, when the first factor consists of 3 levels and the second factor consists of 2 levels, then there will be 3 x 2 = 6 different treatment groups.

As an illustration, let's suppose we are planting corn. The kind of seed and kind of fertilizer are the two factors we are considering in this illustration. This illustration consists of 15 treatment groups. There are.

$3 - 1 = 2$ degrees of freedom for the kind of see and $5 - 1 = 4$ degrees of freedom for the kind of fertilizer. There are $2*4 = 8$ degrees of freedom for the interaction between the kind of see, and the kind of fertilizer.

The data which really appears in the table are samples. In this condition, two samples from each and every treatment group were taken.

	Fert I	Fert II	Fert III	Fert IV	Fert V
Seed A-402	106, 110	95, 100	94, 107	103, 104	100, 102
Seed B-894	110, 112	98, 99	100, 101	108, 112	105, 107
Seed C-952	94, 97	86, 87	98, 99	99, 101	94, 98

Main Effect:

The main effect includes the independent variables one at a time. The interaction is ignored for this portion. Just the rows and columns are utilized, not mixed. This is the portion that is identical to the one-way analysis of variance. Each of the variances computed to analyze the main effects are similar to the between variances.

Interaction Effect:

This can be defined as the effect that one factor has on the other factor. The degree of freedom here is the product of two degrees of freedom for each and every factor.

Within Variation:

This variation is the sum of squares in each and every treatment group. We have one less than the sample size (recall, all the treatment groups should have similar sample sizes for a two-way ANOVA) for each and every treatment group. The total number of treatment groups is the product of a number of levels for each and every factor. Within variance is the within variation divided by its degree of freedom.

The within the group is as well termed as the error.

F-Tests:

There is an F-test for each and every hypothesis, and the F-test is the mean square for each of the main effects and the interaction effect divided by within variance. The numerator degree of freedom comes from each effect, and the denominator degree of freedom is the degree of freedom within variance in each and every case.

Two-Way ANOVA Table:

It is supposed that the main effect A consists of "a" level (and A has a - 1 df, main effect B consists of "b" levels (and B has b - 1 df), n is the sample size of each and every treatment and N = abn is the net sample size.

Note that the overall degree of freedom is once again one less than the net sample size.

Source	SS	df	MS	F
Main Effect A	*given*	A, a-1	SS / df	MS(A) / MS(W)
Main Effect B	*given*	B, b-1	SS / df	MS(B) / MS(W)
Interaction Effect	*given*	A*B, (a-1)(b-1)	SS / df	MS(A*B) / MS(W)
Within	*given*	N - ab, ab(n-1)	SS / df	
Total	sum of others	N - 1, abn - 1		

Summary:

The given outcomes are computed by using the computer. It gives the p-value and critical values for $\alpha = \mathbf{0.05}$.

Source of Variation	SS	df	MS	F	P-value	F-critical
Seed	512.8667	2	256.4333	28.283	0.000008	3.682
Fertilizer	449.4667	4	112.3667	12.393	0.000119	3.056
Interaction	143.1333	8	17.8917	1.973	0.122090	2.641
Within	136.0000	15	9.0667			
Total	1241.4667	29				

From the above outcomes, we can observe that the main effects are both significant; however, the interaction between them is not. That means: the seeds are not all the same & the kinds of fertilizer are not all equal. However, the kind of seed does not interact with the kind of fertilizer. The seeds and fertilizers are independent produce contributing and are additive.

Randomized Complete Block with One Factor

This example illustrates the use of PROC ANOVA in analyzing a randomized complete block design. Researchers are interested in whether three treatments have different effects on the yield and worth of a particular crop. They believe that the experimental units are not homogeneous. So, a blocking factor is introduced that allows the experimental units to be homogeneous within each block. The three treatments are then randomly assigned within each block.

The data from this study are input into the SAS data set *RCB*:

Title1 'Randomized Complete Block'.

data RCB.

input Block Treatment $ Yield Worth @@;

data lines.

1 A 32.6 112   1 B 36.4 130   1 C 29.5 106

2 A 42.7 139   2 B 47.1 143   2 C 32.9 112

3 A 35.3 124   3 B 40.1 134   3 C 33.6 116

The variables *Yield* and *Worth* are continuous response variables, and the variables *Block* and *Treatment* are the classification variables. Because the data for the analysis is balanced, you can use PROC ANOVA to run the analysis.

The SAS statements for the analysis are:

---

proc anova data=RCB;

```
class Block Treatment;

model Yield Worth=Block Treatment;

run;
```

The *Block* and *Treatment* effects appear in the <u>CLASS</u> statement. The <u>MODEL</u> statement requests an analysis for each of the two dependent variables, *Yield,* and *Worth.*

<u>Figure 23.5</u> shows the "Class Level Information" table.

Figure 23.5 Class Level Information

Randomized Complete Block

The ANOVA Procedure

Class Level Information

Class	Levels	Values
Block	3	1 2 3
Treatment	3	A B C

Number of Observations Read 9

Number of Observations Used 9

The "Class Level Information" table lists the number of levels and their values for all effects specified in the <u>CLASS</u> statement. The number of observations in the data set is also displayed. Use this information to make sure that the data have been read correctly.

The overall ANOVA table for *Yield* in <u>Figure 23.6</u> appears first in the output because it is the first response variable listed on the left side in the <u>MODEL</u> statement.

Figure 23.6 Overall ANOVA Table for Yield

Randomized Complete Block

The ANOVA Procedure

Dependent Variable: Yield

Source	DF	Sum of Squares	Mean Square	F Value	Pr > F
Model	4	225.2777778	56.3194444	8.94	0.0283
Error	4	25.1911111	6.2977778		
Corrected Total	8	250.4688889			

R-Square	Coeff Var	Root MSE	Yield Mean
0.899424	6.840047	2.509537	36.68889

The overall F statistic is significant $(F = 8.94, p = 0.0283)$, indicating that the model as a whole account for a significant portion of the variation in *Yield* and that you can proceed to evaluate the tests of effects.

The degrees of freedom (DF) are used to ensure the correctness of the data and model. The Corrected Total degrees of freedom is one less than the total number of observations in the data set;

in this case, $9 - 1 = 8$. The Model degrees of freedom for a randomized complete block is $(b - 1) + (t - 1)$, where the b = number of block levels and the t = number of treatment levels are. In this case, this formula leads to $(3 - 1) + (3 - 1) = 4$ model degrees of freedom.

Several simple statistics follow the ANOVA table. The R-Square indicates that the model accounts for nearly 90% of the variation in the variable *Yield*. The coefficient of variation (C.V.) is listed along with the Root MSE and the mean of the dependent variable. The Root MSE is an estimate of the standard deviation of the dependent variable. The C.V. is a unitless measure of variability.

The tests of the effects shown in <u>Figure 23.7</u> are displayed after the simple statistics.

Figure 23.7 Tests of Effects for Yield

Source	DF	Anova SS	Mean Square	F Value	Pr > F
Block	2	98.1755556	49.0877778	7.79	0.0417
Treatment	2	127.1022222	63.5511111	10.09	0.0274

For *Yield*, both the *Block* and *Treatment* effects are significant $(F = 7.79, p = 0.0417$ and $F = 10.09, p = 0.0274$, respectively) at the 95% level. From this, you can conclude that blocking is useful for this variable and that some contrast between the treatment means is significantly different from zero.

<u>Figure 23.8</u> shows the ANOVA table, simple statistics, and tests of effects for the variable *Worth*.

Figure 23.8 ANOVA Table for Worth

Randomized Complete Block

The ANOVA Procedure

Dependent Variable: Worth

Source	Degrees of freedom	Sum of Squares	Mean Square	F Value	Pr > F
Model	4	1247.33	311.83	8.28	0.0323
Error	4	150.67	37.666667		
Corrected Total	8	1398.00			

R-Square	Coeff Var	Root MSE	Worth Mean
0.892227	4.949450	6.137318	124.0000

Source	DF	Anova SS	Mean Square	F Value	Pr > F
Block	2	354.6666667	177.3333333	4.71	0.0889
Treatment	2	892.6666667	446.3333333	11.85	0.0209

The overall F test is significant $(F = 8.28, p = 0.0323)$ at the 95% level for the variable *Worth*. The *Block* effect is not significant at the 0.05 level but is significant at the 0.10 confidence level $(F = 4.71, p = 0.0889)$. Generally, the usefulness of blocking should be determined before the analysis.

However, since there are two dependent variables of interest, and *Block* is significant for one of them (*Yield*), blocking appears to be generally useful. For *Worth*, as with *Yield*, the effect of *Treatment* is significant $(F = 11.85, p = 0.0209)$.

Issuing the following SAS command produces the *Treatment* means.

---

means Treatment;

run;

[Figure 23.9](#) displays the treatment means and their standard deviations for both dependent variables.

Figure 23.9 Means of Yield and Worth

Randomized Complete Block

The ANOVA Procedure

Level of Treatment	N	Yield		Worth	
		Mean	Std Dev	Mean	Std Dev
A	3	36.8666667	5.22908532	125.000000	13.5277493
B	3	41.2000000	5.43415127	135.666667	6.6583281
C	3	32.0000000	2.19317122	111.333333	5.0332230

Chapter Ten

EXPERIMENTAL DESIGN: Two-Factor Factorial

Greenhouse example (Using SAS)

To put it into perspective, let's look at the phrase 'Experimental Design,' a term you often hear. We are going to take this colloquial phrase and divide it into two formal components:

A) The Treatment Design, and

B) The Randomization Design

We will use the Treatment Design to address the nature of the experimental factors under study and Randomization to address how treatments are assigned to experimental units. An **experimental unit** is defined to be that which receives a treatment level. Simply, this would be a **single plant in the greenhouse** receiving fertilizer. So, the experimental unit here is the plant. (We will discuss later in the course how this is not always quite so clear!)

An ANOVA model often contains elements from both the treatment design and the randomization process, and working with these separately will help us devise the correct ANOVA model for a given experimental situation.

We can imagine **testing three kinds of fertilizers** and also one group one group of untreated plants. The plant biologist kept all the plants under controlled conditions in the greenhouse to focus on the **only uncontrolled** effect of the fertilizer; the only thing we know to differ among the plants. At the end of the experiment, the biologist measured the height of each plant. This **height measurement** is the dependent or **response variable** and is plotted on the vertical (y) axis.

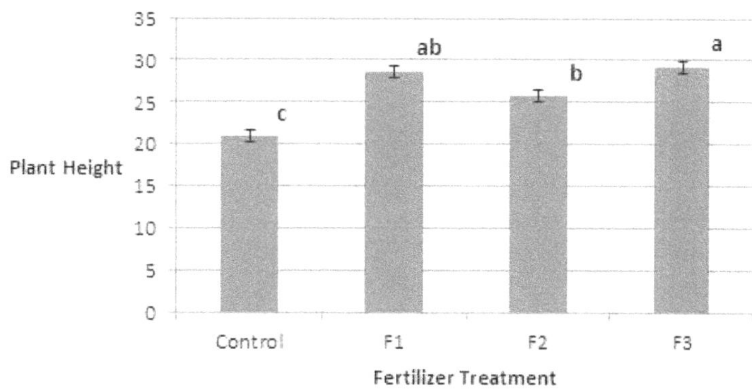

This bar chart is customary to show treatment (or factor) level means. In this case, there was only one treatment: the fertilizer **FACTOR**. This FACTOR, the fertilizer treatment, had four levels: control, F1, F2, and F3, including the control, which received no fertilizer. Using this language convention is important because later on, we will be using ANOVA to handle multi-factor studies (for example, if the biologist manipulated the amount of water AND the type of fertilizer), and we will need to be able to refer to different treatments, each with their own set of levels.

Let us take our greenhouse example further and think about another **FACTOR** the addition of the plant species impacting the plant height. We could have also run a multi-factor experiment to compare two species (Species A and Species B). Our plant (experimental unit) is subjected to 2 factories (Fertilizers and species). We would then need to assign fertilizer and species-level combinations to 48 pots to have six replications in the greenhouse. This would be referred to as a 2 × 4 factorial treatment design.

The data might look like this:

		Control	F1	F2	F3
Species	A	21.0	32.0	22.5	28.0
		19.5	30.5	26.0	27.5
		22.5	25.0	28.0	31.0
		21.5	27.5	27.0	29.5
		20.5	28.0	26.5	30.0
		21.0	28.6	25.2	29.2
	B	23.7	30.1	30.6	36.1
		23.8	28.9	31.1	36.6
		23.8	30.9	28.1	38.7
		23.7	34.4	34.9	37.1
		22.8	32.7	30.1	36.8
		24.4	32.7	25.5	37.1

The ANOVA table will now be constructed as follows:

Source	df	SS	MS	F
Fertilizer	(4 -1) = 3			
Species	(2 -1) = 1			
Fertilizer × Species	(2 -1)(4 -1) = 3			
Error	47 - 7 = 40			
Total	N - 1 = 47			

In SAS, we need first to re-format the data into a 'stacked' format. You can do this in an Excel worksheet and then copy and paste the stacked data into either SAS. Here is an example of the stack data in SAS (greenhouse_2way_sascode.txt).

ALL OF THE FOLLOWING CODE IS EXPLAINED LATER ON

```
data greenhouse_2way;
  input fert $ species $ height;
  datalines;
  control SppA       21.0
  control SppA       19.5
  control SppA       22.5
  control SppA       21.5
  control SppA       20.5
  control SppA       21.0
  control SppB       23.7
  control SppB       23.8
  control SppB       23.8
  control SppB       23.7
  control SppB       22.6
  control SppB       24.4
  f1      SppA       32.0
  f1      SppA       30.5
  f1      SppA       25.0
  f1      SppA       27.5
  f1      SppA       28.0
  f1      SppA       28.6
  f1      SppB       30.1
  f1      SppB       28.9
  f1      SppB       30.9
  f1      SppB       34.4
  f1      SppB       32.7
  f1      SppB       32.7
  f2      SppA       22.5
  f2      SppA       26.0
  f2      SppA       28.0
      ETC.
  f3      SppA       31.0
  f3      SppA       29.5
  f3      SppA       30.0
  f3      SppA       29.2
  f3      SppB       36.1
  f3      SppB       36.6
  f3      SppB       38.7
  f3      SppB       37.1
  f3      SppB       36.8
  f3      SppB       37.1
  ;
run;

proc mixed data=greenhouse_2way method=type3;
  class fert species;
  model height = fert species fert*species;
  store out2way;
run;

ods graphics on;
ods html style=statistical sge=on;
proc plm restore=out2way;
  lsmeans fert*species / adjust=tukey plot=meanplot cl lines;
  /* Because the 2-factor interaction is significant, we need to work with
  the treatment combination means */
  ods exclude diffs diffplot;
run; title; run;
```

The piece of the SAS code above that runs the two-factor factorial model is:

```
proc mixed data=greenhouse_2way method=type3;
class fert species;
model height = fert species fert*species;
store out2way;
run;
```

PROC MIXED is the same SAS procedure we used for the single-factor ANOVA. Again, we specify the data. The method is type 3, which is how the F test is calculated. We specify the two factors as class variables because they are categorical. The model statement should look familiar. Compare this to the "theoretical" illustration in 4.1. Do you see the similarities? The store command stores the elements for our LS interval plot.

As a means of review, think about where the sums of squares are coming from. We have covered the sums of squares for the individual factors in our previous lessons, but now you can see from the output that we have sums of squares for the interaction term. The following (partial) output shows the interaction term in the model:

| Type 3 Analysis of Variance | | | | | | | | |
Source	DF	Sum of Squares	Mean Square	Expected Mean Square	Error Term	Error DF	F Value	Pr > F
fert	3	745.437500	248.479167	Var(Residual) + Q(fert,fert*species)	MS(Residual)	40	73.10	<.0001
species	1	236.740833	236.740833	Var(Residual) + Q(species,fert*species)	MS(Residual)	40	69.65	<.0001
fert*species	3	50.584167	16.861389	Var(Residual) + Q(fert*species)	MS(Residual)	40	4.96	0.0051
Residual	40	135.970000	3.399250	Var(Residual)				

or summarized in the table of Type 3 Tests of Fixed Effects:

| Type 3 Tests of Fixed Effects | | | | |
Effect	Num DF	Den DF	F Value	Pr > F
fert	3	40	73.10	<.0001
species	1	40	69.65	<.0001
fert*species	3	40	4.96	0.0051

So what is our conclusion about this data? Interpret the interaction term first. Is this interaction term significant? Yes! (You can see p-value = 0.0051) So you do NOT interpret the individual factors (fert or species).

***** RULE ***** Interpreting the interaction term first makes sense if you think about it. If there is a significant interaction (as we see above), then it tells us that the effect of the levels of Factor-I depends upon what level of Factor II you are considering. Likewise, in order to evaluate the effect of Factor II, you need to specify what level of Factor-I you are considering. In other words, in the presence of a significant interaction, you can't unambiguously interpret the main effects of treatments involved in the interaction. In the case where an interaction is not significant, we can drop the interaction term from our model (See the next page in this lesson: 4.1.1a -The Additive Model (No Interaction)).

So now that we have looked at the ANOVA output and see the significant interaction term, we know that we want to generate the LS means for the interaction effect (i.e., the treatment combinations) for mean comparisons and plot our figure. The SAS code (from the program greenhouse_2way.sas) that generates these results looks like this:

```
ods graphics on;
ods html style=statistical sge=on;
proc plm restore=out2way;
lsmeans fert*species / adjust=tukey plot=meanplot cl lines;
/* Because the 2-factor interaction is significant, we need to work with
the treatment combination means */
ods exclude diffs diffplot;
run; title; run;
```

SAS Output for the LSmeans:

fert'species Least Squares Means									
fert	species	Estimate	Standard Error	DF	t Value	Pr > |t|	Alpha	Lower	Upper
control	SppA	21.0000	0.7527	40	27.90	<.0001	0.05	19.4788	22.5212
control	SppB	23.7000	0.7527	40	31.49	<.0001	0.05	22.1788	25.2212
f1	SppA	28.6000	0.7527	40	38.00	<.0001	0.05	27.0788	30.1212
f1	SppB	31.6167	0.7527	40	42.00	<.0001	0.05	30.0954	33.1379
f2	SppA	25.8667	0.7527	40	34.37	<.0001	0.05	24.3454	27.3879
f2	SppB	30.0500	0.7527	40	39.92	<.0001	0.05	28.5288	31.5712
f3	SppA	29.2000	0.7527	40	38.79	<.0001	0.05	27.6788	30.7212
f3	SppB	37.0667	0.7527	40	49.25	<.0001	0.05	35.5454	38.5879

Note that the p-values here ($Pr > t$) are testing the hypotheses that the estimates of the means = 0. This is not usually of interest. Notice also that we see a single value for the standard error based on the MSE from the ANOVA, rather than a separate standard error for each mean (as we would get from Proc Summary for the sample means). Again, in this example, with equal sample sizes and no covariates, the LS means will be identical to the ordinary means we would see with the Summary Procedure.

The results of the Tukey test were:

Tukey Grouping for fert'species Least Squares Means (Alpha=0.05)				
LS-means with the same letter are not significantly different.				
fert	species	Estimate		
f3	SppB	37.0667	A	
f1	SppB	31.6167	B	
			B	
f2	SppB	30.0500	B	
			B	
f3	SppA	29.2000	C	B
			C	B
f1	SppA	28.6000	C	B
			C	
f2	SppA	25.8667	C	D
				D
control	SppB	23.7000	E	D
			E	
control	SppA	21.0000	E	

The PLM procedure produces the graph, which we can then label, using the Graphics Editor, with the mean comparison lettering:

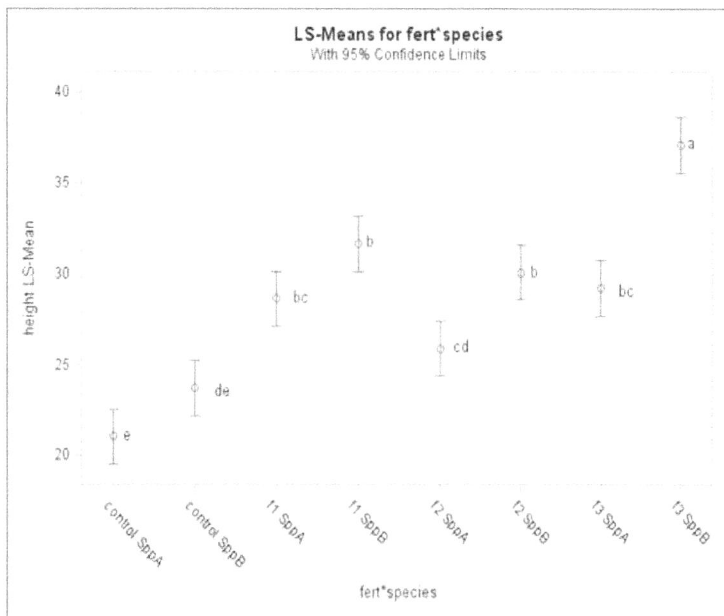

If a traditional bar chart is desired, you can create the graph in Excel by first pasting the LSmeans and standard errors into an Excel spreadsheet, sorting the LSmeans by species, then creating two series: one for SppA, and then another series for SppB. When we plot the means by series, we get a graph of the response (Height) vs. Fertilizer with the second crossed factor (Species) indicated in a legend:

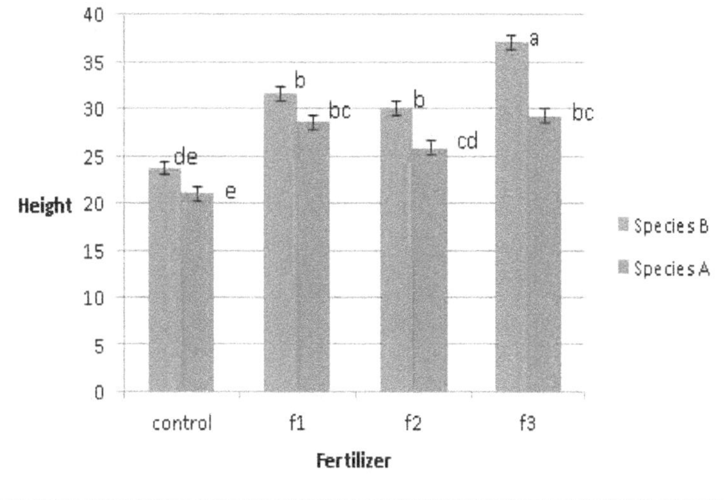

You can see here, as a finished product, we elected to show the standard errors associated with the LSmeans. Because LSmeans can differ from the original sample treatment level means, a safe bet is to plan on using the LSmeans and associated standard errors from the ANOVA for the finished plot.

Using the information from the ANOVA, LSmeans, and mean comparisons to identify differences between treatment levels is an important final step.

APPENDICES

<u>Appendix I</u>: Practice Problems on Regression and ANOVA with answers on the last page

APPENDIX I Problems with answers on Regression

NOTE: These are just Practice Problems. This is NOT meant to look just like the test, and it is NOT the only thing that you should study. Make sure you know all the material from the notes, quizzes, suggested homework

1. The parameters to be estimated in the simple linear regression model $Y = \alpha + \beta x + \varepsilon$ $\varepsilon \sim N(0, \sigma)$ are:
a) α, β, σ b) $\alpha, \beta, \varepsilon$ c) a, b, s   d) $\varepsilon, 0, \sigma$

2. We can measure the proportion of the variation explained by the regression model by:
a) r   b) R^2 c) σ^2 d) F

3. The MSE is an estimator of:
a) ε b) 0   c) σ^2 d) Y

4. In multiple regression with p predictor variables, when constructing a confidence interval for any β_i, the degrees of freedom for the tabulated value of t should be:
a) n-1   b) n-2   c) n-p-1   d) p-1

5. In a regression study, a 95% confidence interval for β_1 was given as: (-5.65, 2.61). What would a test for H_0: $\beta_1 = 0$ vs H_a: $\beta_1 \neq 0$ conclude?
a) reject the null hypothesis at $\alpha = 0.05$ and all smaller α
b) fail to reject the null hypothesis at $\alpha = 0.05$ and all smaller α
c) reject the null hypothesis at $\alpha = 0.05$ and all larger α
d) fail to reject the null hypothesis at $\alpha = 0.05$ and all larger α

6. In simple linear regression, when β is **not** significantly different from zero we conclude that:
a) X is a good predictor of Y   b) there is no linear relationship between X and Y
c) the relationship between X and Y is quadratic   d) there is no relationship between X and Y

7. In a study of the relationship between X=mean daily temperature for the month and Y=monthly charges on electrical bill, the following data was gathered:
Which of the following seems the most likely model?

X	20	30	50	60	80	90
Y	125	110	95	90	110	130

a) $Y = \alpha + \beta x + \varepsilon$ $\beta < 0$
b) $Y = \alpha + \beta x + \varepsilon$ $\beta > 0$
c) $Y = \alpha + \beta_1 x + \beta_2 x^2 + \varepsilon$ $\beta_2 < 0$
d) $Y = \alpha + \beta_1 x + \beta_2 x^2 + \varepsilon$ $\beta_2 > 0$

8. If a predictor variable x is found to be highly significant we would conclude that:
a) a change in y causes a change in x   b) a change in x causes a change in y
c) changes in x are not related to changes in y   d) changes in x are associated to changes in y

9. At the same confidence level, a prediction interval for a new response is always:
a) somewhat larger than the corresponding confidence interval for the mean response
b) somewhat smaller than the corresponding confidence interval for the mean response
c) one unit larger than the corresponding confidence interval for the mean response
d) one unit smaller than the corresponding confidence interval for the mean response

10. Both the prediction interval for a new response and the confidence interval for the mean response are narrower when made for values of x that are:
a) closer to the mean of the x's   b) further from the mean of the x's
c) closer to the mean of the y's   d) further from the mean of the y's

11. In the regression model $Y = \alpha + \beta x + \varepsilon$ the change in Y for a one unit increase in x:
a) will always be the same amount, α b) will always be the same amount, β
c) will depend on the error term   d) will depend on the level of x

12. In a regression model with a dummy variable **without** interaction there can be:
a) more than one slope and more than one intercept   b) more than one slope, but only one intercept
c) only one slope, but more than one intercept   d) only one slope and one intercept

13. In a multiple regression model, where the x's are predictors and y is the response, multicollinearity occurs when:
a) the x's provide redundant information about y
b) the x's provide complementary information about y
c) the x's are used to construct multiple lines, all of which are good predictors of y
d) the x's are used to construct multiple lines, all of which are bad predictors of y

14. Compute the simple linear regression equation if:

	mean	stdev	correlation
x	163.5	16.2	-0.774
y	874.1	54.2	

15. Match the statements below with the corresponding terms from the list.

a) multicollinearity   b) extrapolation
c) R^2 adjusted   d) quadratic regression
e) interaction   f) residual plots
g) fitted equation   h) dummy variables
i) cause and effect   j) multiple regression model
k) R^2 l) residual
m) influential points   n) outliers

_____ Used when a numerical predictor has a curvilinear relationship with the response.

_____ Worst kind of outlier, can totally reverse the direction of association between x and y.

_____ Used to check the assumptions of the regression model.

_____ Used when trying to decide between two models with different numbers of predictors.

_____ Used when the effect of a predictor on the response depends on other predictors.

_____ Proportion of the variability in y explained by the regression model.

_____ Is the observed value of y minus the predicted value of y for the observed x..

_____ A point that lies far away from the rest.

_____ Can give bad predictions if the conditions do not hold outside the observed range of x's.

_____ Can be erroneously assumed in an observational study.

_____ $y = \alpha + \beta_1 x_1 + \beta_2 x_2 + \ldots + \beta_p x_p + \varepsilon \quad \varepsilon \sim N(0, \sigma^2)$

_____ $\hat{y} = a + b_1 x_1 + b_2 x_2 + \ldots + b_p x_p$

_____ Problem that can occur when the information provided by several predictors overlaps.

_____ Used in a regression model to represent categorical variables.

Questions 16 - 19 Palm readers claim to be able to tell how long your life will be by looking at a specific line on your hand. The following is a plot of age of person at death (in years) vs length of life line on the right hand (in cm) for a sample of 28 (dead) people.

16. If we fit a simple linear regression model to these data, what would the value of r be?

a) close to -1

b) close to 0

c) close to 1

d) it's impossible to tell

17. Would you say:

a) length of life line is a very good predictor of age of person at death

b) length of life line is a poor predictor of age of person at death

c) length of life line is a reasonably good predictor of age of person at death

d) cannot determine how good a predictor length of life line is of age of person at death

18. The ANOVA p-value will be around

a) 1.00   b) 0.000   c) 0.05   d) 0.01

19. A better way of modeling age of person at death using this data set would be to use:

a) a nonparametric procedure   c) a contingency table

b) the average age at death   d) quadratic regression

20. According to the null hypothesis of the ANOVA F test, which predictor variables are providing significant information about the response?

a) most of them   b) none of them   c) all of them   d) some of them

21. According to the alternative hypothesis of the ANOVA F test, which predictor variables are providing significant information about the response?

a) most of them   b) none of them   c) all of them   d) some of them

22. In general, the Least Squares Regression approach finds the equation:

a) that includes the best set of predictor variables

b) of the best fitting straight line through a set of points

c) with the highest R^2, after comparing all possible models

d) that has the smallest sum of squared errors

23. Studies have shown a high positive correlation between the number of firefighters dispatched to combat a fire and the financial damages resulting from it. A politician commented that the fire chief should stop sending so many firefighters since they are clearly destroying the place. This is an example of:

a) extrapolation

b) dummy variables

c) misuse of causality

d) multicollinearity

24. The following appeared in the magazine *Financial Times*, March 23, 1995: "When Elvis Presley died in 1977, there were 48 professional Elvis impersonators. Today there are an estimated 7328. If that growth is projected, by the year 2012 one person in four on the face of the globe will be an Elvis impersonator." This is an example of:

a) extrapolation

b) dummy variables

c) misuse of causality

d) multicollinearity

Questions 25 – 43 Most supermarkets use scanners at the checkout counters. The data collected this way can be used to evaluate the effect of price and store's promotional activities on the sales of any product. The promotions at a store change weekly, and are mainly of two types: flyers distributed outside the store and through newspapers (which may or may not include that particular product), and in-store displays at the end of an aisle that call the customers' attention to the product. Weekly data was collected on a particular beverage brand, including sales (in number of units), price (in dollars), flyer (1 if product appeared that week, 0 if it didn't) and display (1 if a special display of the product was used that week, 0 if it wasn't).

As a preliminary analysis, a simple linear regression model was done.
The fitted regression equation was: sales = 2259 - 1418 price.
The ANOVA F test p-value was .000, and R^2= 59.7%.

25. The response variable is:
a) quantitative   b) y   c) sales   d) all of the above

26. Which of the following is the best interpretation of the slope of the line?
a) As the price increases by 1 dollar, sales will increase, on average, by 2259 units.
b) As the price increases by 1 dollar, sales will decrease, on average, by 1418 units.
c) As the sales increase by 1 unit, the price will increase, on average, by 2259 dollars.
d) As the sales increase by 1 unit, the price will decrease, on average, by 1418 dollars.

27. Should the intercept of the line be interpreted in this case?
a) Yes, as the average price when no units are sold.
b) Yes, as the average sales when the price is zero dollars.
c) No, since sales of zero units are probably out of the range observed.
d) No, since a price of zero dollars is probably out of the range observed.

28. The proportion of the variability in sales accounted for by the price of the product is:
a) 14.18%   b) 22.59%
c) 59.70%   d) 100%

29. The coefficient of linear correlation, r, for this analysis is:
a) 7.73   b) -7.73
c) .773   d) -.773

30. According to this model, how many units will be sold, on average, when the price of the beverage is $1.10?
a) 3818.8
b) 699.2
c) 1066.9
d) 3902.9

31. Is price a good predictor of sales?
a) Yes, the p-value is very small.   b) Yes, the intercept is very large.
c) No, R-square is not too good.   d) No, the slope is negative.

32. Below is a sketch of the residual plot for this analysis. What can you conclude from it?
a) All the assumptions seem to be satisfied.
b) There seems to be an outlier in the data.
c) Simple linear regression might not be the best model.
d) The assumption of constant variance might be violated.

Next, a quadratic regression was fitted to the data. Parts of the computer output appear below.

Predictor	Coef	Stdev	t-ratio	p
Constant	7990.0	724.7	11.03	0.000
price	-10660	1151	-9.26	0.000
price2	3522.3	436.8	____	0.000

Analysis of Variance

SOURCE	DF	SS	MS	F	p
Regression	2	16060569	8030284	125.11	0.000
Error	60	3851231	64187		
Total	62	19911800			

33. We can write the model fitted here as:

a) $Y = \alpha + \beta x + \varepsilon$ b) $Y = \alpha + \beta_1 x_1 + \beta_2 x_2 + \varepsilon$

c) $Y = \alpha + \beta_1 x + \beta_2 x^2 + \varepsilon$ d) $Y = \alpha + \beta_1 x_1 + \beta_2 x_2 + \beta_3 x_1 x_2 + \varepsilon$

34. What is R^2?

a) 19.3%   b) 80.7%   c) 23.98%   d) 76.06%

35. What is the test statistic to determine if the quadratic term significantly differs from zero?

a) 125.11   b) 11.03   c) -9.26   d) 8.06

36. Based on the results of the two regression analyses presented here, which of the following sketches best describes the relationship between price of the item and sales?

a)  b)  c)  d) 

37. According to this model, how many units will be sold, on average, when the price of the beverage is $1.10?

a) 525.98
b) 138.53
c) 10660
d) 3522.3

38. Is the quadratic model preferable to the linear model in this case?

a) No, we always prefer the simpler model.
b) No, the p-value for the quadratic term is zero.
c) Yes, the p-value for the quadratic term is zero.
d) Yes, we had more data for the quadratic model.

Next, the categorical variables flyer and display were added to the model. Parts of the computer output appear below.

Predictor	Coef	Stdev	t-ratio	p
Constant	3829.5	700.4	5.47	0.000
price	-5056	1026	-4.93	0.000
price2	1667.7	369.9	4.51	0.000
flyer	804.12	86.75	9.27	0.000
display	-31.49	53.38	-0.59	0.558

$s = 162.8$ R-sq = 92.3%   R-sq(adj) = 91.7%

Analysis of Variance

SOURCE	DF	SS	MS	F	p
Regression	4	18373664	4593416	173.21	0.000
Error	58	1538137	26520		
Total	62	19911802			

39. According to this model, what is the average effect of advertising the product on the weekly flyer, after adjusting for price and display?
a) There will be a significant increase in sales of about 804 units.
b) There will be an insignificant increase in sales of about 9 units.
c) Since the effect is significant, we should not interpret the coefficient.
d) Since the effect is not significant, we should not interpret the coefficient.

40. According to this model, what is the average effect of promoting the product with an in-store display, after adjusting for price and flyer?
a) There will be an insignificant increase in sales of less than one unit.
b) There will be a significant decrease in sales of about 31 units.
c) Since the effect is significant, we should not interpret the coefficient.
d) Since the effect is not significant, we should not interpret the coefficient.

41. What can you say about price and price squared in this model?
a) Both of them are still good predictors of sales.
b) Neither of them seems to be a good predictor of sales now.
c) Price is a good predictor, but price squared is not.
d) Price is not a good predictor, but price squared is.

42. Should anything be done to improve this model?
a) No, it has very good ANOVA p-value, R-sq and R-sq adjusted.
b) No, it has a lot of parameters so it does a good job of predicting sales.
c) Yes, not all the variables included in the model are good predictors of sales.
d) Yes, "price" and "display" should be taken out since they have negative coefficients.

43. Flyers usually advertise products that are on sale that week. This implies that we should add to the model:
a) an interaction between display and flyer.   b) an interaction between price and flyer.
c) a quadratic term for display.   d) a quadratic term for flyer.

Questions 44 – 50 A scientific foundation wanted to evaluate the relation between y= salary of researcher (in thousands of dollars), x_1= number of years of experience, x_2= an index of publication quality, x_3=sex (M=1, F=0) and x_4= an index of success in obtaining grant support. A sample of 35 randomly selected researchers was used to fit the multiple regression model. Parts of the computer output appear below.

Predictor	Coef	SE Coef	T	P
Constant	17.846931	2.001876	8.915	0.0001
Years	1.103130	0.359573	3.068	0.0032
Papers	0.321520	0.037109	0.0002	
Sex	1.593400	0.687724	2.317	0.0083
Grants	1.288941	0.298479	4.318	0.0003

$s = 1.75276$ R-sq = 92.3%   adj R-sq = 91.4%

44. The least squares line fitted to the data is:
a) salary = $2.001 + 0.33 x_1 + 0.04 x_2 + 0.69 x_3 + 0.30 x_4 + \varepsilon$
b) salary = $17.85 + 1.10 x_1 + 0.32 x_2 + 1.59 x_3 + 1.29 x_4 + \varepsilon$
c) salary = $2.001 + 0.33 x_1 + 0.04 x_2 + 0.69 x_3 + 0.30 x_4$
d) salary = $17.85 + 1.10 x_1 + 0.32 x_2 + 1.59 x_3 + 1.29 x_4$

45. The p-value of the ANOVA F test will be:
a) very large, since it's clear that all of the variables are good predictors of salary
b) very small, since it's clear that all of the variables are good predictors of salary
c) very large, since it's clear that none of the variables are good predictors of salary
d) very small, since it's clear that none of the variables are good predictors of salary

46. The (one-sided) p-value for testing whether salary increases with years of experience is:
a) .0001   b) .0032   c) .0064   d) .0016

47. The variable that helps the most in predicting salary is:
a) intercept   b) years   c) papers   d) sex   e) grants

48. Which of the following gives a 95% CI for β_1?
a) $17.847 \pm t^* (2.002)$ b) $17.847 \pm t^* (8.915)$ c) $1.1031 \pm t^* (.3596)$ d) $1.1031 + t^* (3.068)$

49. How many degrees of freedom does the t^* value from the previous question have?
a) 34   b) 33   c) 30   d) 4

50. According to the assumptions, what has to have a Normal distribution and constant variance?
a) the researchers   b) the years   c) the variables   d) the salaries

Questions 51 – 53 In a study on teenage pregnancies, the researchers attempted to determine the relationship between y=weight of baby at birth (in pounds) and x=age of the mother. The data collected is plotted below.

51. If we fit a simple linear regression model to these data, the value of r will be closest to?
a) 0   b) -1   c) 1   d) 100

52. To determine if age of mother is a significantly good predictor of weight of baby we could:
a) construct a confidence interval for the slope
b) test whether the slope is zero or not
c) use an ANOVA F test
d) all of the above

```
wt 9  -       •  •  •
      -    •   •   •   •
      -  •  •  •  •          •
7     -  • •  •       •   •
      -  • •      •  •  •
      -    •  •  •  •  •
5     -  •  • •  • •
         - - - - - - - - - -   age of
         12     15     18      mother
```

53. The best fitting line through these points will probably have a slope that is:
a) positive and significantly different from zero.
b) positive but not significantly different from zero.
c) negative and significantly different from zero.
d) negative but not significantly different from zero.

Questions 54 – 63 The following is a plot of the same data, but using different symbols to represent mothers who received prenatal care (*) and those who didn't (•). We fit the model:
$Y = \alpha + \beta_1 x_1 + \beta_2 x_2 + \beta_3 x_1 x_2 + \varepsilon$, where Y= wt x_1=age x_2= 0 for prenatal care, 1 if not

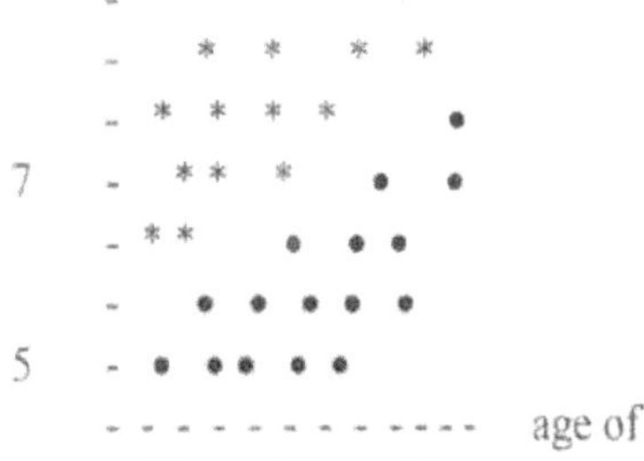

54. The model we fit to these data is <u>not</u>:
a) Multiple Regression
b) Regression with Dummy Variables
c) Least Squares Regression
d) Quadratic Regression

55. The baseline group is:
a) age   b) weight   c) prenatal care   d) random

Match each parameter with the interpretations at right:
56. β_3 a) change in intercept
57. β_2 b) intercept for baseline group
58. β_1 c) slope for baseline group
59. α d) change in slope
                       e) slope for non-baseline group

60. For the model above, and based on the plot, which of the following statements is true?
a) Age of mother does not seem to be a good predictor of weight of baby.
b) Prenatal care will probably have a significant effect on weight of baby.
c) There appears to be an interaction between prenatal care and age of mother.
d) all the statements are false.

61. Which parameter would we test for to determine if the rate at which weight of baby increases with age of mother differs for mothers who receive prenatal care and those who don't?
a) β_3 b) β_2 c) β_1 d) α

The computer output reports following equation: wt= -1.84 + 0.53x_1 + 1.79x_2 - .003 $x_1 x_2$.
Use that equation to answer the next two questions:

62. To predict weight of babies born to teenagers who do <u>not</u> receive prenatal care we use the equation:
a) wt= -0.05 + .527x_1 b) wt= -1.84 + 0.53x_1
c) wt= 1.79 + .003x_1 d) wt= 1.79 + 0.53x_1

63. To predict weight of babies born to teenagers who receive prenatal care we use the equation:
a) wt= -0.05 + .527x_1 b) wt= -1.84 + 0.53x_1
c) wt= 1.79 + .003x_1 d) wt= 1.79 + 0.53x_1

Questions 64 -70 Data for 51 U.S. "states" (50 states, plus the District of Columbia) was used to examine the relationship between violent crime rate (violent crimes per 100,000 persons per year) and the predictor variables of urbanization (percentage of the population living in urban areas) and poverty rate. A predictor variable indicating whether or not a state is classified as a Southern state (1 = Southern, 0 = not) was also included. Some Minitab output for the analysis of this data is shown below (with some information intentionally left blank).

```
The regression equation is
Crime = -321.9 +4.69Urban +39.3Poverty -649.3South +12.1Urban*South -5.84Poverty*South

Predictor         Coef    SE Coef      T       P
Constant       -321.90     148.20   -2.17   0.035
Urban             4.689      1.654    2.83   0.007
Poverty          39.34      13.52    2.91   0.006
South(S=1)     -649.30     266.96   -2.43   0.019
Urban*South      12.05       2.871    4.20   0.000
Poverty*South    -5.838     16.671   -0.35   0.728

Analysis of Variance
Source            DF        SS       MS        F       P
Regression         5   2060459   412091    ------   0.000
Residual Error    45    882169    19604
Total             50   2942628
```

64. Which of the following represents the fitted relationship between crime, urbanization, and poverty for <u>Southern</u> states?
a) Crime = –321.9 + 4.69Urban + 39.3Poverty
b) Crime = –315.6 + 4.69Urban + 39.3Poverty
c) Crime = –315.6 + 16.8Urban + 33.5Poverty
d) Crime = –971.2 + 16.8Urban + 33.5Poverty
e) Crime = –971.2 + 4.69Urban + 39.3Poverty

65. Predict the violent crime rate for a Southern state with an urbanization of 55.4 and a poverty rate of 13.7.
a) 417.2   b) 510.1   c) 535.8
d) 582.4   e) 633.5

66. Predict the violent crime rate for a non-Southern state with an urbanization of 65.6 and a poverty rate of 8.0.
a) 300.4   b) 336.5   c) 349.1
d) 416.9   e) 432.2

67. Calculate the ANOVA F test statistic value.
a) 2.34   b) 4.20   c) 4.58
d) 21.02   e) 47.00

68. When finding the p-value for the ANOVA F test, what degrees of freedom should be used?
a) df = 5   b) df = 45
c) df = 50   d) $df_1 = 5$, $df_2 = 45$ e) $df_1 = 5$, $df_2 = 50$

69. Based on the p-value for the ANOVA F test shown in the output, how many of the predictors are useful for predicting crime rate?
a) none of them   b) all of them   c) exactly one of them   d) at least one of them

70. Which of the following predictors should probably be removed from the model to improve it?
a) Urban   b) Poverty   c) South   d) Urban*South   e) Poverty*South

Questions 71 - 79 The National Math and Science Initiative (NMSI) has recently begun a controversial program in which high school students are paid cash incentives for passing an end-of-year standardized test. Suppose we conduct a similar study, in which end-of-year test scores (y) are measured on a scale of 0–100 and the amount of the cash incentive offered to the student (x) is measured in dollars from $0 to $500. A scatterplot of the 96 observations in the sample and the regression line is shown below, along with some Minitab output (with some information intentionally left blank).

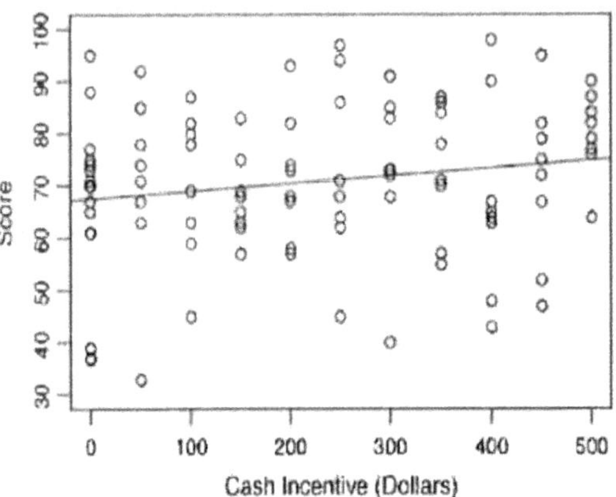


```
The regression equation is
Score = 67.51 +0.0148Cash

Predictor     Coef   SE Coef      T       P
Constant    67.513    2.509    ——    ——
Cash       0.014762  0.00886    ——    ——

Predicted Values for New Observations
Cash    Fit  SE Fit    95% CI     95% PI
200     ——    ——      —[A]—      —[B]—

Cash    Fit  SE Fit    95% CI     95% PI
500     ——    ——      —[C]—      —[D]—
```

71. The coefficient 0.01476 in the equation is

a) the parameter α.

b) the parameter β.

c) our estimate of the parameter α.

d) our estimate of the parameter β.

72. Which of the following is the best interpretation of the slope?

a) For each additional dollar offered to students, their scores increase by 0.01476 points, on average.

b) For each additional dollar offered to students, their scores increase by 67.51 points, on average.

c) For each 0.01476 additional dollars offered to students, their scores increase by one point, on average.

d) For each 67.51 additional dollars offered to students, their scores increase by one point, on average.

73. Which of the following is the best interpretation of the intercept?

a) The predicted score of a student who is offered no cash is 0.01476 points.

b) The predicted score of a student who is offered no cash is 67.51 points.

c) The amount of cash offered to a student who scores a zero is, on average, 0.01476 dollars.

d) The amount of cash offered to a student who scores a zero is, on average, 67.51 dollars.

e) None of the above; it is not appropriate to interpret the intercept in this situation.

74. Calculate the predicted test score of a student who is offered a cash incentive of $200.

a) 64.56   b) 70.46   c) 73.41   d) 76.37   e) 82.27

75. The p-value for the ANOVA test was 0.0952, so there is _______ evidence that scores on the test depend on the size of the cash incentive.

a) not enough   b) pretty strong   c) very strong   d) some   e) no

76. Calculate the value of the t test statistic for testing whether score depends on cash incentive.

a) 0.600   b) 1.67   c) 2.79   d) 4.10   e) 9.59

77. Which of the four intervals labeled as [A]–[D] in the Minitab output would be the *widest*?

a) [A]   b) [B]   c) [C]   d) [D]   e) All four would have the same width.

78. Which of the four intervals labeled as [A]–[D] in the Minitab output would be the *narrowest*?

a) [A]   b) [B]   c) [C]   d) [D]   e) All four would have the same width.

79. Which of the four intervals labeled as [A]–[D] in the Minitab output would be the confidence interval for β?

a) [A]   b) [B]   c) [C]   d) [D]   e) None of these

Questions 80 - 87 The economic structure of Major League Baseball allows some teams to make substantially more money than others, which in turn allows some teams to spend much more on player salaries. These teams might therefore be expected to have better players and win more games on the field as a result. Suppose that after collecting data on team payroll (in millions of dollars) and season win total for 2010, we find a regression equation of Wins = 71.87 + 0.101Payroll - 0.060League where League is an indicator variable that equals 0 if the team plays in the National League or 1 if the team plays in the American League.

80. If Teams A and B both play in the same league, and Team A's payroll is $1 million higher than Team B's, then we would expect Team A to win, on average,
a) 0.101 games more than Team B.
b) 71.87 games more than Team B.
c) 0.060 games more than Team B.
d) 0.060 games fewer than Team B.

81. If Teams A and B have the same payroll, but Team A plays in the National League while Team B plays in the American League, then we would expect Team A to win, on average,
a) 0.101 games more than Team B.
b) 71.87 games more than Team B.
c) 0.060 games more than Team B.
d) 0.060 games fewer than Team B.

82. Suppose we plotted the data and drew the regression lines for National League and American League teams. What would be the slope of the line for American League teams?
a) −0.060　　b) 0.060　　c) 0.941
d) 0.101　　e) 71.81

83. Suppose we plotted the data and drew the regression lines for National League and American League teams. What would be the intercept of the line for American League teams?
a) −0.060　　b) 0.060　　c) 0.941
d) 0.101　　e) 71.81

84. Calculate the predicted number of wins for a National League team with a payroll of $98 million.
a) 65.99　　b) 77.75　　c) 77.85
d) 81.71　　e) 81.77

85. One American League team in the data set had a payroll of $108 million and won 88 games. Calculate the residual for this observation.
a) −1.26　　b) 5.28　　c) 9.65
d) 11.70　　e) 22.61

86. The t tests for which variable would have the same p-value as the ANOVA test?
a) constant　　b) payroll
c) league　　d) wins
e) none of them

87. Common sense suggests that teams with a higher payroll should have a strong tendency to win more games, but that league affiliation should not matter. Then common sense suggests that the ANOVA F test for this data would probably have
a) a small test statistic value and a small p-value.
b) a small test statistic value and a large p-value.
c) a large test statistic value and a small p-value.
d) a large test statistic value and a large p-value.

88. Based on the common sense described in the previous question, in which of the following t tests would we probably reject the null hypothesis?
a) the t test for Payroll　　b) the t test for League
c) both t tests　　d) neither t test

Questions 89 -95 Ecologists have long known that there is a relationship between the amount of precipitation a location receives and the number of trees that grow in the area. Suppose that the yearly rainfall (x, measured in mm) and the amount of the ground covered by trees (y, measured on a scale from 0 to 100) are recorded for 49 geographic locations. In the sample data, x has a sample mean of 1182.4 and a sample standard deviation of 226.0, while y has a sample mean of 49.6 and a sample standard deviation of 7.1. The sample correlation between x and y is 0.673.

89. In a simple linear regression analysis of this data, when we write $y = \alpha + \beta x + \varepsilon$, which of the following do we assume?
a) The x values are independent and normally distributed with mean 0 and constant variance.
b) The x values are independent and normally distributed with variance 0 and constant mean.
c) The errors are independent and normally distributed with mean 0 and constant variance.
d) The errors are independent and normally distributed with variance 0 and constant mean.
e) both a) and c)

90. Use the information provided to calculate the regression equation.
a) TreeCover = 24.70 + 0.0211 Rainfall
b) TreeCover = 0.0211 + 24.70 Rainfall
c) TreeCover = –25371.8 + 21.5 Rainfall
d) TreeCover = 21.5 – 25371.8 Rainfall
e) TreeCover = 25471.0 + 21.5 Rainfall

91. Calculate the predicted amount of tree cover for an area that receives 1230 mm of rainfall per year.
a) 50.6   b) 52.3   c) 55.9   d) 60.9   e) 63.8

92. What percentage of the variability in tree cover is explained by rainfall?
a) 2.1%   b) 21.5%   c) 24.7%   d) 45.3%   e) 67.3%

93. For this data set, find the degrees of freedom for regression.
a) 1   b) 2   c) 47   d) 48   e) 49

94. For this data set, find the degrees of freedom for error.
a) 1   b) 2   c) 47   d) 48   e) 49

95. In a regression t test for this data, which of the following statements is the *alternative* hypothesis (in words)?
a) The population mean of tree cover is not zero.
b) The population mean of tree cover is zero.
c) Tree cover depends on rainfall.
d) Tree cover does not depend on rainfall.
e) The population means of tree cover and rainfall are not equal.

96. Suppose now that we record a 50th observation, include it in the data set, and recalculate the regression equation. Which of the following possibilities for the 50th observation would probably change the regression equation the most?
a) $x = 900$, $y = 45$
b) $x = 1200$, $y = 40$
c) $x = 1200$, $y = 60$
d) $x = 2400$, $y = 75$
e) $x = 2400$, $y = 25$

ANSWERS

1. a
2. b
3. c
4. c
5. b
6. b
7. d (Plot the data)
8. d
9. a
10. a
11. b
12. c
13. a
14. y-hat = 1297.49 – 2.59 x
15. d m f c e k l n b I j g a h
16. b
17. b
18. a
19. b
20. b
21. d
22. d
23. c
24. a
25. d
26. b
27. d
28. c
29. d
30. b
31. a
32. c
33. c
34. b
35. d
36. d
37. a
38. c
39. a
40. d
41. a
42. c
43. b
44. d
45. b
46. d
47. c
48. c
49. c
50. d

51. a
52. d
53. b
54. d
55. c
56. d
57. a
58. c
59. b
60. b
61. a
62. a
63. b
64. d
65. a
66. a
67. d
68. d
69. d
70. e
71. d
72. a
73. b
74. b
75. d
76. b
77. d
78. a
79. e
80. a (Write down the two equations
81. c   and plot them for 80 – 88)
82. d
83. e
84. e
85. b
86. e
87. c
88. a
89. c
90. a
91. a
92. d
93. a
94. c
95. c
96. e

APPENDIX II: SAS Least Squares Linear Regression

Once it is determined from the scatterplot that the relationship is linear with a constant variance, you will likely want to fit the least squares regression line. The least squares regression line is the line that minimizes the sum of the squares of the residuals. The regression equation is $y=\beta_0+\beta_1 x$.

The procedure we use is PROC REG. The format of PROC REG is,

PROC REG data=library.datafile;

MODEL *the model (see below)*;

RUN ;

The *model* is the Regression model which you are testing; for example, if you were testing *price* as a function of *sqfeet* (size), the model would be *price=sqfeet*. This means that we are testing whether or the ot *price* is dependent on *sqfeet*.

Example

Suppose we wanted to see if sqfeet affects house price.

The SAS program we would write is,

PROC REG data=sasuser.houses;

MODEL *price=sqfeet.*

RUN;

SAS will output the results in the OUTPUT wind; it should look something like the output in the following window.,

A Remark: At this junction, we can add that the only difference between the simple and multiple regression in the SAS program is the fact that the MODEL will have many variables on the right side of the "=" symbol; and also the many options that can be selected in the analysis

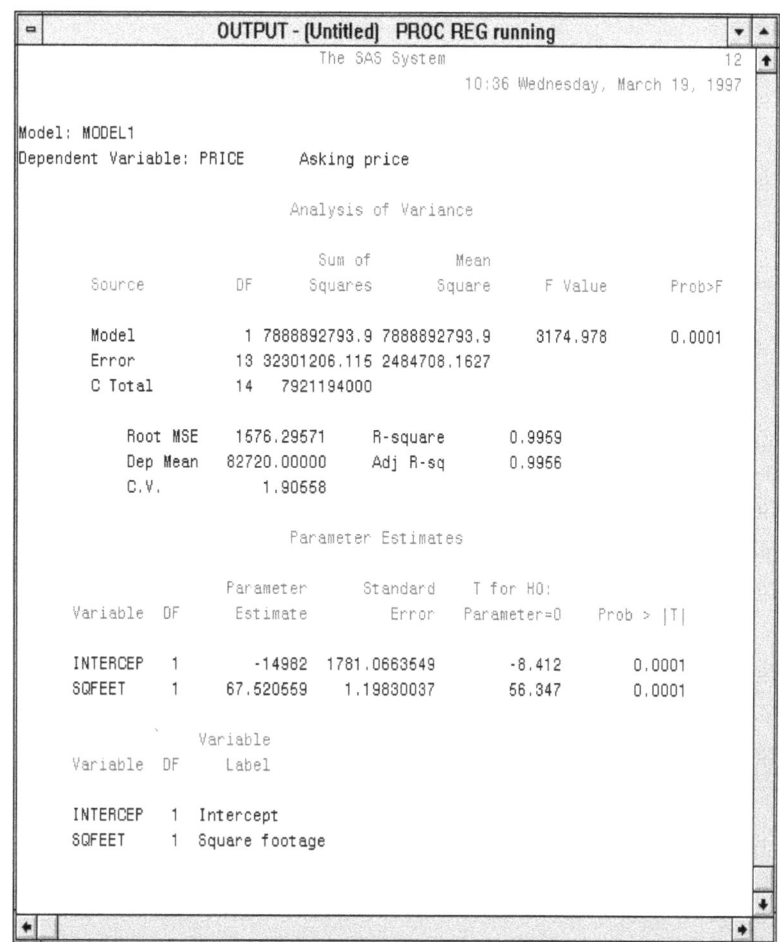

From the output, we can see that the regression is significant ($p < 0.000{,}1$) and we can see from the parameter estimates that the fitted line will be:

price = -14982 + 67.5206 * *sqfeet*.

So, for every additional square foot of house area, we would expect to pay an additional $67.52.

Multicollinearity in Multiple Regression

Multicollinearity in regression occurs when predictor variables (independent variables) in the regression model are more highly correlated with other predictor variables than with the dependent variable. Multicollinearity does not adversely affect the regression equation if the purpose of your research is only to predict the dependent variable from a set of predictor variables. In this case, the predictions in your regression will still be accurate, and the overall $R2$ will give you an indication of how well the predictor variables in your model predict the dependent variable. Multicollinearity does not affect the goodness of fit and the goodness of prediction.

In regression, multicollinearity can be a problem if the purpose of your study is to estimate the contributions of individual predictors. When multicollinearity is present, p values can be misleading, and the regression coefficients' confidence intervals will be very wide and may vary dramatically with the addition or exclusion of just one case/participant. If this is the case, removing any highly correlated terms from the model will greatly affect the estimated coefficients of the other highly correlated terms. Multicollinearity inflates parameter variances ($\beta_0, \beta_1, \ldots, \beta_p$) estimates. This may lead to lack of statistical significance of individual independent variables even though the overall model may be significant. This is especially true for small and moderate sample sizes. Such problems will result in incorrect conclusions about relationships between independent and dependent variables. This is a mistake you don't want to make in your study.

Now using the height data given above and using formula (a) for the computation, the following calculations will be done:

```
Obs  x   x-mean(x)  (x-mean(x))²

     ---------------------------------

1   175    4.67       21.81

2   183   12.67      160.53

3   165   -5.33       28.41

4   170   -0.33        0.11

5   165   -5.33       28.41

6   193   22.67      513.93

7   183   12.67      160.53

8   175    4.67       21.81
```

9 175   4.67   21.81

10 163 -7.33   53.73

11 150 -20.33   413.31

12 147 -23.33   544.29

N(x)=12

SUM(x)=2044

MEAN(x)=2044/12=170.33

SUM((x-mean(x))²)=1968.68

$$s = \sqrt{\frac{\sum (x - \bar{x})^2}{n-1}} = \sqrt{\frac{1968.68}{12-1}} = 13.38$$

Hence the Standard Deviation is 13.38. Note that in the example had calculations rounded to a low precision - when using a software package like SAS, the precision used is far higher, so the accuracy will be better. The same result will come out if the Standard Deviation is calculated using formulas (b) or (c), but the sum of the observation values may exceed what a calculator may allow, particularly if the values are large.

SAS Code

As stated at the start of this paper, these statistics can be calculated using the SAS UNIVARIATE procedure, although the MEANS, SUMMARY, and other procedures that produce the same statistics can also be used. The SAS code needed is:

proc univariate;

var height;

run;

The output from this program is shown below:

The UNIVARIATE Procedure

Variable: height

Moments

N   12   Sum Weights   12

Mean   170.333333   Sum Observations   2044

Std Deviation   13.3779556   Variance   178.969697

Skewness   -0.2639558   Kurtosis   -0.153035

Uncorrected SS   350130   Corrected SS   1968.66667

Coeff Variation   7.8539857   Std Error Mean   3.86188314

Basic Statistical Measures

Location   Variability

Mean   170.3333   Std Deviation   13.37796

Median 172.5000   Variance   178.96970

Mode   175.0000   Range   46.00000

Interquartile Range   15.00000

Tests for Location: Mu0=0

Test   -Statistic-   -----p Value------

Student's t   t 44.10629   Pr > |t| <.0001

Sign   M   6   Pr >= |M| 0.0005

Signed Rank   S   39   Pr >= |S| 0.0005

Quantiles (Definition 5)

Quantile   Estimate

100% Max   193.0

99%   193.0

95%   193.0

90%   183.0

75% Q3   179.0

50% Median   172.5

25% Q1   164.0

10%   150.0

5%   147.0

1%   147.0

0% Min   147.0

Extreme Observations

Lowest----   ----Highest---

Value   Obs   Value   Obs

147   12   175   8

150   11   175   9

163	10	183	2
165	5	183	7
165	3	193	6

From the output, the statistics discussed above can be retrieved. The only statistic that may be hard to find is the Minimum and Maximum values, but these can be found in the Quantiles section of the output. In this case, the Minimum value is 147, and the Maximum value is 193.

APPENDIX III: TESTING HYPOTHESES on MEANS

SINGLE SAMPLE

The first part of this paper will look at the procedure where there is a single sample and uses the sample mean to test the hypothesis. The process for the test is summarized below:

1. The first step is to state the null and alternative hypothesis. As the test on the mean the question being asked, for a two-tailed test, is

$H_0: \mu = \mu_0$

$H_1: \mu \neq \mu_0$

For a single tailed test the alternative hypothesis would be written as

$H_1: \mu > \mu_0$

or

$H_1: \mu < \mu_0$

depending on the question being asked.

2. The next step in the procedure is to decide what the level of significance should be, usually noted as α in most textbooks, and calculates what the boundaries of the critical region on the t curve are. If a two-sided test is required, then the α value is split evenly between the two sides of the t-curve. To find the critical t-values, it is necessary to consult a t-test statistical table, an example of which is given below:

   t-Distribution

1)   significance

2)   Degrees   --------------------------------------

3)	of Freedom	0.100	0.050	0.025	0.010	0.005
4)		-----	-----	-----	-----	-----
5)	1	3.078	6.314	12.706	31.821	63.657
6)	2	1.886	2.920	4.303	6.965	9.925
7)	3	1.638	2.353	3.182	4.541	5.841
8)	4	1.533	2.132	2.776	3.747	4.604
9)	5	1.476	2.015	2.571	3.365	4.032
10)	6	1.440	1.943	2.447	3.143	3.707
11)	7	1.415	1.895	2.365	2.998	3.499
12)	8	1.397	1.860	2.306	2.896	3.355
13)	9	1.383	1.833	2.262	2.821	3.250
14)	10	1.372	1.812	2.228	2.764	3.169
15)	11	1.363	1.796	2.201	2.718	3.106
16)	12	1.356	1.782	2.179	2.681	3.055
17)	13	1.350	1.771	2.160	2.650	3.012
18)	14	1.345	1.761	2.145	2.624	2.977
19)	15	1.341	1.753	2.131	2.602	2.947
20)	16	1.337	1.746	2.120	2.583	2.921
21)	17	1.333	1.740	2.110	2.567	2.898
22)	18	1.330	1.734	2.101	2.552	2.878

23)   19   1.328 1.729 2.093 2.539 2.861

24)   20   1.325 1.725 2.086 2.528 2.845

25)   21   1.323 1.721 2.080 2.518 2.831

26)   22   1.321 1.717 2.074 2.508 2.819

27)   23   1.319 1.714 2.069 2.500 2.807

28)   24   1.318 1.711 2.064 2.492 2.797

29)   25   1.316 1.708 2.060 2.485 2.787

30)   26   1.315 1.706 2.056 2.479 2.779

31)   27   1.314 1.703 2.052 2.473 2.771

32)   28   1.313 1.701 2.048 2.467 2.763

33)   29   1.311 1.699 2.045 2.462 2.756

34)   1.310 1.697 2.042 2.457 2.750

Before the correct t-value can be obtained it is necessary to decide on the "degrees of freedom", which is equal to n-1 - if there are 15 observations, then the degrees of freedom that are consulted on a table is 15-1=14. If the test is a two-sided and uses an α level of 0.10 then the boundary is $t_{0.05}=1.761$ and $-t_{0.05}=-1.761$.

3. The only real computation to do in this test is calculate a value for t from the data using the following formula:

$$t = \frac{\bar{x} - u_0}{s/\sqrt{n}} \quad (1.1)$$

4. With the critical boundaries known and a value of t computed from the formula above it is now decision time - the hypothesis H_1 is rejected if the value t is in the critical region; otherwise, accept the hypothesis H_0.

The following examples demonstrate the procedure used for the tests.

Example 1

A sample of eight bottles of a certain product was taken, and their liquid content was measured – the results are below:

369, 357, 356, 364, 348, 361, 345, 364

The researcher wants to test the null hypothesis that the mean equals 355 versus the alternative that it does not. Let $\alpha = 0.01$.

5. $H_0: \mu = 355$, $H_1: \mu \neq 355$

6. As this is a two-tailed test, and $\alpha=0.01$ with (8-1)=7 degrees of freedom then from the t-table $t_{0.005}=3.499$, $-t_{0.005}=-3.499$.

7. Do computations.

$$n=8$$

$$\bar{x}=\frac{\sum x}{n}=\frac{2864}{8}=358$$

$$S=\sqrt{\frac{\sum x^2-n\bar{x}^2}{n-1}}=\sqrt{\frac{1025788-8\times358^2}{7}}=8.25$$

$$t=\frac{358-355}{8.25/\sqrt{8}}=1.029$$

8   As -3.499 < 1.029 < 3.499 then accept H_0.

Now looking at SAS, how does the same test get done? There is a procedure called PROC TTEST that will do the calculations; however, the default output and interpretation are very different. Using the same data as in example 1 (loading it into a dataset called PRODA with a variable VOLUME) and running the following code.

20   proc ttest data=proda;

21   var volume;

22   run;

will produce the following output.

The TTEST Procedure

Statistics

Variable	N	Lower CL Mean	Mean	Upper CL Mean	Lower CL Std Dev	Std Dev	Upper CL Std Dev	Std Err
volume	8	346.25	358	369.75	4.6472	8.2462	24.477	2.9155

T-Tests

Variable	DF	t Value	Pr > |t|
volume	7	1.03	0.3377

From the output, how is a decision made as to whether to accept or reject the hypothesis? The number to look for is under the label "Pr > $|t|$," and to read this correctly, the number is compared against the significance level - if the number is greater than or equal to the significance then accept H_0, otherwise, accept H_1. In this case, the significance was 0.05 (0.10/2=0.05 since the two-sided test), and as 0.3,377 > 0.005, then accept H_0.

Is there a way to check that the p-value from the TTEST procedure is acceptable or that the correct hypothesis was chosen? After all the way the procedure for determining which hypothesis to choose is well-known and in countless textbooks? There is a function available in BASE SAS that can give the p-value from the t value computed in Example 1 and is shown in the SAS data step below with the result:

31   data _null_;

32   x=1.029;

33   df=7;

34   p=(1-probt(abs(x),df))*2; /*significance level of a two-tailed t test*/

35   put p=;

36   run;

p=0.3377168268

The t-test can also be done using the UNIVARIATE procedure for the Single Sample case and using the MU0= option as the following SAS code and output shows (look at the Tests for Location section, Student's t result):

393   proc univariate data=proda mu0=355;

394   var volume;

395   run;

The UNIVARIATE Procedure

Variable: Volume

Moments

N   8 Sum Weights   8

Mean   358 Sum Observations   2864

Std Deviation   8.24621125 Variance   68

Skewness   -0.480995 Kurtosis   -0.7557588

Uncorrected SS   1025788 Corrected SS   476

Coeff Variation   2.30341096 Std Error Mean   2.91547595

Basic Statistical Measures

Location   Variability

Mean   358.0000   Std Deviation   8.24621

Median 359.0000   Variance   68.00000

Mode   364.0000   Range   24.00000

Interquartile Range   12.00000

Tests for Location: Mu0=355

Test   -Statistic-   -----p Value------

Student's t   t 1.028992   Pr > |t| 0.3377

Sign   M   2   Pr >= |M| 0.2891

Signed Rank   S   7   Pr >= |S| 0.3672

It is also possible to do the calculations using data step code within BASE SAS, as shown below, and get an output similar to the output below::

```
data _null_;

SAS Statements...

tsigl=-abs(tinv(alpha,df));

tsigh=abs(tinv(alpha,df));

tval=(mean-mju)/(std/sqrt(n));

p=(1-probt(abs(tval),df))*2;

more SAS Statements...

run;
```

--- Output ---

T-TEST

Dataset = PRODA

Variable = volume

H0 = 355 , H1 ^= 355

alpha = 0.01 (2-sided test: alpha/2=0.005)

N= 8

Mean=358

S= 8.2462112512

DF= 7

3.499483297 < 1.0289915109 < 3.4994832974 : accept H0, reject H1

Pr > |t| = 0.3377205477

For the data step method I have as a macro in my macro collection – every programmer should carry around with them something that contains their useful of frequently used code.

Example 2

A company took a random sample of ten components and clocked the duration a machine took to recondition and inspect the each component (in seconds), the times of which were

5.7, 4.8, 5.9, 4.9, 6.1, 4.2, 6.5, 6.4, 5.8, 5.7

The goal is to have an average time of 5 seconds. Using a significance level of 0.01 was the goal met?

8. H_0: $\mu = 5$, H_1: $\mu > 5$ (only concerned if the average time is greater than 5 seconds)

9. As this is a single-tailed test and $\alpha=0.01$ with $(10-1)=9$ degrees of freedom then from the t-table $t_{0.01}=2.821$.

10. Computation

$$n=10$$

$$\bar{x}=5.6$$

$$S=0.74$$

$$t=\frac{5.6-5}{0.74/\sqrt{10}}=2.564$$

11. As $2.564<2.821$ then accept H_o.

In the previous example, had the significance level been 0.05 ($t_{0.05}=1.833$) then the result would have been quite different, $1.833<=2.564$ then reject H_o and accept H_1.

Using SAS and the UNIVARIATE procedure (MJU=5) the output is:

The UNIVARIATE Procedure

Variable: time

Moments

N	10	Sum Weights	10
Mean	5.6	Sum Observations	56
Std Deviation	0.74087036	Variance	0.54888889
Skewness	-0.7500206	Kurtosis	-0.2612091
Uncorrected SS	318.54	Corrected SS	4.94
Coeff Variation	13.2298278	Std Error Mean	0.23428378

Basic Statistical Measures

Location   Variability

Mean   5.600000   Std Deviation   0.74087

Median 5.750000   Variance   0.54889

Mode   5.700000   Range   2.30000

Interquartile Range   1.20000

Tests for Location: Mu0=5

Test   -Statistic-   -----p Value------

Student's t   t 2.560997   Pr > |t| 0.0306

Sign   M   2   Pr >= |M| 0.3438

Signed Rank   S   19   Pr >= |S| 0.0488

A decision is made as to accept H_0 or H_1 by comparing the "Pr > |t|" value against the significance level – in this case 0.01 < 0.0306 so accept H_0.

TWO SAMPLES

The second part of this paper will discuss using the *t*-test to compare two means with an unknown but assumed common population variance.

The hypothesis that is usually tested is that the means are equal, denoted by H_0: $\mu_1=\mu_2$. The hypothesis can also be rewritten as $H_0:\mu_1-\mu_2=0$ - this is useful as it is then possible to easily write the test to check if it is larger or smaller by a specified value, sometimes denoted in textbooks as Δ.

The test is very much the same as for the single sample except that the calculation for *t* is now:

$$t = \frac{(\bar{x}_1 - \bar{x}_2) - (\bar{u}_1 - \bar{u}_2)}{\sqrt{s_p^2 \left(\frac{1}{n_1} + \frac{1}{n_2}\right)}} \qquad (2.1)$$

where

$$s_p^2 = \frac{(n_1 - 1)s_1^2 + (n_2 - 1)s_2^2}{n_1 + n_2 - 2} = \frac{\sum (x_{1i} - \bar{x}_1)^2 + \sum (x_{2i} - \bar{x}_2)^2}{n_1 + n_2 - 2} \qquad (2.2)$$

Summarizing the procedure the process would be:

12. First step is to state the null and alternative hypothesis. As the test on the mean the question being asked, for a two-tailed test, is

H_0: $\mu_1 - \mu_2 = \Delta$

H_1: $\mu_1 - \mu_2 \neq \Delta$

For a single tailed test the alternative hypothesis would be written as

H_1: $\mu_1 - \mu_2 > \Delta$

or

H_1: $\mu_1 - \mu_2 < \Delta$

depending on the question being asked.

13. The next step in the procedure is to decide what the level of significance should be as above.

Before the correct t-value can be obtained it is necessary to decide on the "degrees of freedom", which is equal to $n_1 + n_2 - 2$ - if there are 7 observations in sample 1, 8 observations in sample 2, then the degrees of freedom that is consulted on a table is 7+8-2=13. If the test is a two-sided and uses an α level of 0.10 then the boundary is $t_{0.05} = 1.771$ and $-t_{0.05} = -1.771$.

14. The only real computation to do in this test is calculate a value for t from the data using calculations 2.1 and 2.2.

15. With the critical boundaries known and a value of t computed from the formula above it is now decision time - the hypothesis H_1 is rejected if the value t is in the critical region; otherwise, accept the hypothesis H_0.

The following examples demonstrate the procedure used for the tests.

Example 3

A sample of free range eggs from two farms were sought and a test was asked for to determine if the mean weight (ounces) of the eggs from the two farms were the same using a significance of 0.01:

Farm A: 20, 28, 24, 20, 24, 21, 17, 28, 25, 19

Farm B: 29, 16, 25, 27, 27, 18, 22, 27

16. H_0: $\mu_1 - \mu_2 = 0$, H_1: $\mu_1 - \mu_2 \neq 0$

17. As this is a two-tailed test, and $\alpha=0.01$ with $(10+8-2)=16$ degrees of freedom then from the t-table $t_{0.005}=2.921$, $-t_{0.005}=-2.921$.

18. Do computations

$$n_1=10, n_2=8$$

$$\bar{x}_1=\frac{\sum x}{n}=\frac{226}{10}=22.6, \bar{x}_2=\frac{191}{8}=23.875$$

$$s_1^2=\frac{\sum x^2-n\bar{x}^2}{n-1}=\frac{5236-10 \times 22.6^2}{10-1}=14.27, s_2^2=\frac{4717-8 \times 23.875^2}{8-1}=22.41$$

$$s_p^2=\frac{(n_1-1)s_1^2+(n_2-1)s_2^2}{n_1+n_2-2}=\frac{(10-1)14.27+(8-1)22.41}{10+8-2}=17.83$$

$$t = \frac{(\bar{x}_1 - \bar{x}_2) - (\bar{u}_1 - \bar{u}_2)}{\sqrt{s_p^2\left(\frac{1}{n_1} + \frac{1}{n_2}\right)}} = \frac{(22.6 - 23.875) - (0)}{\sqrt{17.83\left(\frac{1}{10} + \frac{1}{8}\right)}} = \frac{-1.275}{2.003} = -0.637$$

19. As -2.921 < -0.637 < 2.921 then accept H₀.

To calculate the *t*-test using SAS procedures it is not possible to use the BASE SAS procedures but instead use others, for example TTEST from the SAS/STAT module. Using the data above and using the variable FARM to indicate where the sample came from, WTOZ as the weight in ounces, and the following code:

```
proc ttest data=eggs0;

class farm;

var wtoz;

run;
```

The following output appears:

The TTEST Procedure

Statistics

Lower CL   Upper CL Lower CL   Upper CL

Variable	farm	N	Mean	Mean	Mean	Std Dev	Std Dev	Std Dev	Std Err
wtoz	1	10	19.898	22.6	25.302	2.598	3.7771	6.8956	1.1944
wtoz	2	8	19.917	23.875	27.833	3.13	4.734	9.635	1.6737
wtoz	Diff (1-2)		-5.521	-1.275	2.971	3.1448	4.2225	6.4264	2.0029

T-Tests

Variable	Method	Variances	DF	t Value	Pr > |t|

| wtoz | Pooled | Equal | 16 | -0.64 | 0.5334 |

| wtoz | Satterthwaite | Unequal | 13.3 | -0.62 | 0.5457 |

Equality of Variances

Variable	Method	Num DF	Den DF	F Value	Pr > F
wtoz	Folded F	7	9	1.57	0.5173

As with the UNIVARIATE procedure above the result to look at is the " Pr> |t|" value in the row using the method Pooled (if the assumption is that the two populations have the same variance then the "Pooled" method value is used, otherwise the Satterthwaite method value is used – the distinction is too advanced for this paper and if the reader is interested they should refer to the SAS documentation). As with the single sample method the decision is made as to accept H_0 or H_1 comparing the "Pr > |t|" value against the significance level – in this case $0.005 < 0.5334$ $(0.01/2=0.005$ – two tailed test) so accept H_0.

To check if the p-value calculated in the TTEST procedure is the same as the one that was calculated by hand above the following data step code is used:

```
41  data _null_;

42     x=-0.637;

43     df=16;

44     p=(1-probt(abs(x),df))*2; /*significance level of a two-tailed t test*/

45     put p=;

46  run;

p=0.533133971
```

The value 0.5331 is about that of 0.5334 from the TTEST procedure – the difference is due to rounding.

A good programmer will carry around a piece of SAS code to do this test using BASE SAS, that is the manual method plus the calculation for the p-value.

Example 4

A sample of free range eggs from two farms were sought and a test was asked for to determine if the mean weight (ounces) of the eggs from Farm A was greater than Farm B by three ounces using a significance of 0.01:

Farm A: 26, 26, 28, 33, 32, 27, 24, 24

Farm B: 19, 16, 26, 18, 28, 20, 18, 23, 18, 27

20. H_0: $\mu_1 - \mu_2 = 3$, H_1: $\mu_1 - \mu_2 > 3$

21. As this is a one-tailed test, and $\alpha=0.01$ with $(10+8-2)=16$ degrees of freedom then from the t-table $t_{0.01}=2.583$.

22. Do computations

$$n_1=10, n_2=8$$

$$\bar{x}_1=\frac{\sum x}{n}=\frac{220}{8}=27.5, \bar{x}_2=\frac{213}{10}=21.3$$

$$s_1^2=\frac{\sum x^2-n\bar{x}^2}{n-1}=\frac{6130-8 \times 27.5^2}{8-1}=11.43, s_2^2=\frac{4707-10 \times 21.3^2}{10-1}=18.90$$

$$s_p^2=\frac{(n_1-1)s_1^2+(n_2-1)s_2^2}{n_1+n_2-2}=\frac{(8-1)11.43+(10-1)18.9}{10+8-2}=15.63$$

$$t=\frac{(\bar{x}_1-\bar{x}_2)-(\bar{u}_1-\bar{u}_2)}{\sqrt{s_p^2(\frac{1}{n_1}+\frac{1}{n_2})}}=\frac{(27.5-21.3)-(3)}{\sqrt{15.63(\frac{1}{8}+\frac{1}{10})}}=\frac{3.2}{1.86}=1.72$$

23. As $1.72 < 2.583$ then accept H_0.

Using the TTEST procedure and with the following SAS code

proc ttest data=eggs0 H0=3;

class farm;

var wtoz;

run;

the following output is generated:

The TTEST Procedure

Statistics

Variable	farm	N	Lower CL Mean	Mean	Upper CL Mean	Lower CL Std Dev	Std Dev	Upper CL Std Dev	Std Err
wtoz	1	8	23.317	27.5	31.683	1.9863	3.3806	8.9927	1.1952
wtoz	2	10	16.832	21.3	25.768	2.6853	4.3474	9.9017	1.3748
wtoz	Diff (1-2)		0.7224	6.2	11.678	2.7016	3.9536	6.974	1.8754

T-Tests

Variable	Method	Variances	DF	t Value	Pr > |t|
wtoz	Pooled	Equal	16	1.71	0.1073
wtoz	Satterthwaite	Unequal	16	1.76	0.0981

Equality of Variances

Variable	Method	Num DF	Den DF	F Value	Pr > F
wtoz	Folded F	9	7	1.65	0.5198

A decision is made as to accept H_0 or H_1 by comparing the "Pr > |t|" value against the significance level – in this case $0.005 < 0.1073$ so accept H_0.

SINGLE SAMPLE

The first part of this paper will look at the procedure where there is a single sample and uses the sample mean to test the hypothesis. The process for the test is summarized below:

1. The first step is to state the null and alternative hypothesis. As the test on the mean the question being asked, for a two-tailed test, is

H_0: $\mu = \mu_0$

H_1: $\mu \neq \mu_0$

For a single tailed test, the alternative hypothesis would be written as

H_1: $\mu > \mu_0$

or

H_1: $\mu < \mu_0$

depending on the question being asked.

2. The next step in the procedure is to decide what the level of significance should be, usually noted as α in most textbooks, and calculates what the boundaries of the critical region on the t curve are. If a two-sided test is required, then the α value is split evenly between the two sides of the t-curve. To find the critical t-value, it is necessary to consult a t-test statistical table, an example of which is given below:

3. t-Distribution

4. significance

5. Degrees   -------------------------------------

6. of Freedom 0.100 0.050 0.025 0.010 0.005

7.   ---

8.   1   3.078 6.314 12.706 31.821 63.657

9.   2   1.886 2.920 4.303 6.965 9.925

10.   3   1.638 2.353 3.182 4.541 5.841

11.   4   1.533 2.132 2.776 3.747 4.604

12.   5   1.476 2.015 2.571 3.365 4.032

13.   6   1.440 1.943 2.447 3.143 3.707

14.   7   1.415 1.895 2.365 2.998 3.499

15.   8   1.397 1.860 2.306 2.896 3.355

16.   9   1.383 1.833 2.262 2.821 3.250

17.   10   1.372 1.812 2.228 2.764 3.169

18.   11   1.363 1.796 2.201 2.718 3.106

19.   12   1.356 1.782 2.179 2.681 3.055

20.   13   1.350 1.771 2.160 2.650 3.012

21.   14   1.345 1.761 2.145 2.624 2.977

22.   15   1.341 1.753 2.131 2.602 2.947

23.   16   1.337 1.746 2.120 2.583 2.921

24.   17   1.333 1.740 2.110 2.567 2.898

25.   18   1.330 1.734 2.101 2.552 2.878

26.   19   1.328 1.729 2.093 2.539 2.861

27.	20	1.325	1.725	2.086	2.528	2.845
28.	21	1.323	1.721	2.080	2.518	2.831
29.	22	1.321	1.717	2.074	2.508	2.819
30.	23	1.319	1.714	2.069	2.500	2.807
31.	24	1.318	1.711	2.064	2.492	2.797
32.	25	1.316	1.708	2.060	2.485	2.787
33.	26	1.315	1.706	2.056	2.479	2.779
34.	27	1.314	1.703	2.052	2.473	2.771
35.	28	1.313	1.701	2.048	2.467	2.763
36.	29	1.311	1.699	2.045	2.462	2.756
	30	1.310	1.697	2.042	2.457	2.750

Before the correct t-value can be obtained it is necessary to decide on the "degrees of freedom", which is equal to n-1 - if there are 15 observations, then the degrees of freedom that are consulted on a table is 15-1=14. If the test is a two-sided and uses an α level of 0.10 then the boundary is $t_{0.05}$=1.761 and $-t_{0.05}$=-1.761.

3. The only real computation to do in this test is calculate a value for *t* from the data using the following formula:

$$t = \frac{\bar{x} - u_0}{s/\sqrt{n}} \quad (1.1)$$

4. With the critical boundaries known and a value of t computed from the formula above it is now decision time - the hypothesis H_1 is rejected if the value *t* is in the critical region; otherwise, accept the hypothesis H_0.

The following examples demonstrate the procedure used for the tests.

Example 1

A sample of eight bottles of a certain product was taken, and their liquid content was measured – the results are below:

369, 357, 356, 364, 348, 361, 345, 364

The researcher wants to test the null hypothesis that the mean equals 355 versus the alternative that it does not. Let α = 0.01.

5. H_0: μ = 355, H_1: $\mu \neq$ 355

6. As this is a two-tailed test, and α=0.01 with (8-1)=7 degrees of freedom then from the t-table $t_{0.005}$=3.499, $-t_{0.005}$=-3.499.

7. Do computations.

$n = 8$

$$\bar{x} = \frac{\sum x}{n} = \frac{2864}{8} = 358$$

$$S = \sqrt{\frac{\sum x^2 - n\bar{x}^2}{n-1}} = \sqrt{\frac{1025788 - 8 \times 358^2}{7}} = 8.25$$

$$t = \frac{358 - 355}{8.25/\sqrt{8}} = 1.029$$

8. As -3.499 < 1.029 < 3.499 then accept H_0.

Now looking at SAS, how does the same test get done? There is a procedure called PROC TTEST that will do the calculations; however, the default output and interpretation are very different. Using the same data as in example 1 (loading it into a dataset called PRODA with a variable VOLUME) and running the following code.

20   proc ttest data=proda;

21   var volume;

22   run;

will produce the following output.

The TTEST Procedure

Statistics

		Lower CL		Upper CL	Lower CL		Upper CL	
Variable	N	Mean	Mean	Mean	Std Dev	Std Dev	Std Dev	Std Err
volume	8	346.25	358	369.75	4.6472	8.2462	24.477	2.9155

T-Tests

Variable	DF	t Value	Pr > |t|
volume	7	1.03	0.3377

From the output, how is a decision made as to whether accept or reject the hypothesis? The number to look for is under the label "Pr > |t|," and to read this correctly, the number is compared against the significance level - if the number is greater than or equal to the significance, then accept H_0; otherwise, accept H_1. In this case, the significance was 0.05 (0.10/2=0.05 since two-sided test), and as 0.3377 > 0.005, then accept H_0.

Is there a way to check that the p-value from the TTEST procedure is acceptable or that the correct hypothesis was chosen? After all the way the procedure for determining which hypothesis to choose is well-known and in countless textbooks? There is a function available in BASE SAS that can give the p-value from the t-value computed in Example 1 and is shown in the SAS data step below with the result:

31   data _null_;

32   x=1.029;

33   df=7;

34   p=(1-probt(abs(x),df))*2; /*significance level of a two-tailed t test*/

35   put p=;

36 run;

p=0.3377168268

The t-test can also be done using the UNIVARIATE procedure for the Single Sample case and using the MU0= option as the following SAS code and output shows (look at the Tests for Location section, Student's t result):

393 proc univariate data=proda mu0=355;

394   var volume;

395 run;

The UNIVARIATE Procedure

Variable: Volume

Moments

N	8	Sum Weights	8
Mean	358	Sum Observations	2864
Std Deviation	8.24621125	Variance	68
Skewness	-0.480995	Kurtosis	-0.7557588
Uncorrected SS	1025788	Corrected SS	476
Coeff Variation	2.30341096	Std Error Mean	2.91547595

Basic Statistical Measures

Location   Variability

Mean   358.0000   Std Deviation   8.24621

Median 359.0000   Variance   68.00000

Mode   364.0000   Range   24.00000

Interquartile Range   12.00000

Tests for Location: Mu0=355

Test   -Statistic-   -----p Value------

Student's t   t 1.028992   Pr > |t| 0.3377

Sign   M   2   Pr >= |M| 0.2891

Signed Rank   S   7   Pr >= |S| 0.3672

It is also possible to do the calculations using data step code within BASE SAS, as shown below, and get an output similar to the output below::

```
data _null_;

SAS Statements...

tsigl=-abs(tinv(alpha,df));

tsigh=abs(tinv(alpha,df));

tval=(mean-mju)/(std/sqrt(n));

p=(1-probt(abs(tval),df))*2;

more SAS Statements...

run;
```

Output ---

T-TEST

Dataset = PRODA

Variable = volume

H0 = 355 , H1 ^= 355

alpha = 0.01 (2-sided test: alpha/2=0.005)

N= 8

Mean=358

S= 8.2462112512

DF= 7

 3.499483297 < 1.0289915109 < 3.4994832974 : accept H0, reject H1

Pr > |t| = 0.3377205477

For the data step method, I have as a macro in my macro collection – every programmer should carry around with them something that contains their useful or frequently used code.

Example 2

A company took a random sample of ten components and clocked the duration a machine took to recondition and inspect each component (in seconds), the times of which were

5.7, 4.8, 5.9, 4.9, 6.1, 4.2, 6.5, 6.4, 5.8, 5.7

The goal is to have an average time of 5 seconds. Using a significance level of 0.01, was the goal met?

1. H_0: $\mu = 5$, H_1: $\mu > 5$ (only concerned if the average time is greater than 5 seconds)

2. As this is a single-tailed test and $\alpha=0.01$ with $(10-1)=9$ degrees of freedom then from the t-table $t_{0.01}=2.821$.

3. Computation

$n=10$

$\bar{x}=5.6$

$S=0.74$

$$t=\frac{5.6-5}{0.74/\sqrt{10}}=2.564$$

9. As $2.564<2.821$ then accept H_0.

In the previous example, had the significance level been 0.05 ($t_{0.05}=1.833$), then the result would have been quite different, $1.833<=2.564$, then reject H_0 and accept H_1.

Using SAS and the UNIVARIATE procedure (MJU=5), the output is:

The UNIVARIATE Procedure

Variable: time

Moments

N	10	Sum Weights	10
Mean	5.6	Sum Observations	56
Std Deviation	0.74087036	Variance	0.54888889
Skewness	-0.7500206	Kurtosis	-0.2612091
Uncorrected SS	318.54	Corrected SS	4.94
Coeff Variation	13.2298278	Std Error Mean	0.23428378

Basic Statistical Measures

Location   Variability

Mean   5.600000   Std Deviation   0.74087

Median 5.750000   Variance   0.54889

Mode   5.700000   Range   2.30000

Interquartile Range   1.20000

Tests for Location: Mu0=5

Test   -Statistic-   -----p Value------

Student's t   t 2.560997   Pr > |t| 0.0306

Sign   M   2   Pr >= |M| 0.3438

Signed Rank   S   19   Pr >= |S| 0.0488

A decision is made as to accept H_0 or H_1 by comparing the "Pr > |t|" value against the significance level – in this case 0.01 < 0.0306 so accept H_0.

TWO SAMPLES

The second part of this paper will discuss using the *t*-test to compare two means with an unknown but assumed common population variance.

The hypothesis that is usually tested is that the means are equal, denoted by $H_0: \mu_1 = \mu_2$. The hypothesis can also be rewritten as $H_0: \mu_1 - \mu_2 = 0$ - this is useful as it is then possible to easily write the test to check if it is larger or smaller by a specified value, sometimes denoted in textbooks as Δ.

The test is very much the same as for the single sample except that the calculation for *t* is now:

$$t = \frac{(\bar{x}_1 - \bar{x}_2) - (\bar{u}_1 - \bar{u}_2)}{\sqrt{s_p^2 \left(\frac{1}{n_1} + \frac{1}{n_2}\right)}} \qquad (2.1)$$

where

$$s_p^2 = \frac{(n_1 - 1)s_1^2 + (n_2 - 1)s_2^2}{n_1 + n_2 - 2} = \frac{\sum (x_{1i} - \bar{x}_1)^2 + \sum (x_{2i} - \bar{x}_2)^2}{n_1 + n_2 - 2} \qquad (2.2)$$

Summarizing the procedure, the process would be:

10. First step is to state the null and alternative hypotheses. As the test on the mean, the question being asked for a two-tailed test is

H_0: $\mu_1 - \mu_2 = \Delta$

H_1: $\mu_1 - \mu_2 \neq \Delta$

For a single-tailed test, the alternative hypothesis would be written as

H_1: $\mu_1 - \mu_2 > \Delta$

or

H_1: $\mu_1 - \mu_2 < \Delta$

depending on the question being asked.

11. The next step in the procedure is to decide what the level of significance should be, as above.

Before the correct t-value can be obtained, it is necessary to decide on the "degrees of freedom," which is equal to $n_1 + n_2 - 2$ - if there are seven observations in sample 1 and 8 observations in sample 2, then the degrees of freedom that is consulted on a table is 7+8-2=13. If the test is a two-sided and uses an α level of 0.10, then the boundary is $t_{0.05}=1.771$ and $-t_{0.05}=-1.771$.

12. The only real computation to do in this test is to calculate a value for t from the data using calculations 2.1 and 2.2.

13. With the critical boundaries known and a value of t computed from the formula above, it is now decision time -hypothesis H_1 is rejected if the value t is in the critical region; otherwise, accept the hypothesis H_0.

The following examples demonstrate the procedure used for the tests.

Example 3

A sample of free range eggs from two farms were sought and a test was asked for to determine if the mean weight (ounces) of the eggs from the two farms were the same using a significance of 0.01:

Farm A: 20, 28, 24, 20, 24, 21, 17, 28, 25, 19

Farm B: 29, 16, 25, 27, 27, 18, 22, 27

14. H_0: $\mu_1 - \mu_2 = 0$, H_1: $\mu_1 - \mu_2 \neq 0$

15. As this is a two-tailed test, and $\alpha=0.01$ with $(10+8-2)=16$ degrees of freedom then from the t-table $t_{0.005}=2.921$, $-t_{0.005}=-2.921$.

16. Do computations

$$n_1=10, n_2=8$$

$$\bar{x}_1=\frac{\sum x}{n}=\frac{226}{10}=22.6, \bar{x}_2=\frac{191}{8}=23.875$$

$$s_1^2=\frac{\sum x^2-n\bar{x}^2}{n-1}=\frac{5236-10 \times 22.6^2}{10-1}=14.27, s_2^2=\frac{4717-8 \times 23.875^2}{8-1}=22.41$$

$$s_p^2=\frac{(n_1-1)s_1^2+(n_2-1)s_2^2}{n_1+n_2-2}=\frac{(10-1)14.27+(8-1)22.41}{10+8-2}=17.83$$

$$t = \frac{(\bar{x}_1 - \bar{x}_2) - (\bar{u}_1 - \bar{u}_2)}{\sqrt{s_p^2 \left(\frac{1}{n_1} + \frac{1}{n_2}\right)}} = \frac{(22.6 - 23.875) - (0)}{\sqrt{17.83 \left(\frac{1}{10} + \frac{1}{8}\right)}} = \frac{-1.275}{2.003} = -0.637$$

17. As -2.921 < -0.637 < 2.921 then accept H_o.

To calculate the *t*-test using SAS procedures, it is not possible to use the BASE SAS procedures but instead use others, for example TTEST from the SAS/STAT module. Using the data above and using the variable FARM to indicate where the sample came from, WTOZ as the weight in ounces, and the following code:

```
proc ttest data=eggs0;

class farm;

var wtoz;

run;
```

The following output appears:

The TTEST Procedure

Statistics

			Lower CL		Upper CL	Lower CL		Upper CL	
Variable	farm	N	Mean	Mean	Mean	Std Dev	Std Dev	Std Dev	Std Err
wtoz	1	10	19.898	22.6	25.302	2.598	3.7771	6.8956	1.1944
wtoz	2	8	19.917	23.875	27.833	3.13	4.734	9.635	1.6737
wtoz	Diff (1-2)		-5.521	-1.275	2.971	3.1448	4.2225	6.4264	2.0029

T-Tests

Variable	Method	Variances	DF	t Value	Pr > |t|

| wtoz | Pooled | Equal | 16 | -0.64 | 0.5334 |

| wtoz | Satterthwaite | Unequal | 13.3 | -0.62 | 0.5457 |

Equality of Variances

Variable	Method	Num DF	Den DF	F Value	Pr > F
wtoz	Folded F	7	9	1.57	0.5173

As with the UNIVARIATE procedure above, the result to look at is the " Pr> |t|" value in the row using the method Pooled (if the assumption is that the two populations have the same variance, then the "Pooled" method value is used, otherwise the Satterthwaite method value is used – the distinction is too advanced for this paper and if the reader is interested they should refer to the SAS documentation). As with the single sample method, the decision is made as to accept H_0 or H_1 comparing the "Pr > |t|" value against the significance level – in this case $0.005 < 0.5334$ $(0.01/2=0.005$ – two-tailed test) so accept H_0.

To check if the p-value calculated in the TTEST procedure is the same as the one that was calculated by hand above, the following data step code is used:

```
41  data _null_;

42    x=-0.637;

43    df=16;

44    p=(1-probt(abs(x),df))*2; /*significance level of a two-tailed t test*/

45    put p=;

46  run;

p=0.533133971
```

The value 0.5331 is about that of 0.5334 from the TTEST procedure – the difference is due to rounding.

A good programmer will carry around a piece of SAS code to do this test using BASE SAS, which is the manual method, plus the calculation for the p-value.

Example 4

A sample of free-range eggs from two farms was sought, and a test was asked to determine if the mean weight (ounces) of the eggs from Farm A was greater than Farm B by three ounces using a significance of 0.01:

Farm A: 26, 26, 28, 33, 32, 27, 24, 24

Farm B: 19, 16, 26, 18, 28, 20, 18, 23, 18, 27

18. H_0: $\mu_1 - \mu_2 = 3$, H_1: $\mu_1 - \mu_2 > 3$

19. As this is a one-tailed test, and $\alpha=0.01$ with $(10+8-2)=16$ degrees of freedom then from the t-table $t_{0.01}=2.583$.

20. Do computations

$$n_1=10, n_2=8$$

$$\bar{x}_1=\frac{\sum x}{n}=\frac{220}{8}=27.5, \bar{x}_2=\frac{213}{10}=21.3$$

$$s_1{}^2=\frac{\sum x^2-n\bar{x}^2}{n-1}=\frac{6130-8 \times 27.5^2}{8-1}=11.43, s_2{}^2=\frac{4707-10 \times 21.3^2}{10-1}=18.90$$

$$s_p{}^2=\frac{(n_1-1)s_1{}^2+(n_2-1)s_2{}^2}{n_1+n_2-2}=\frac{(8-1)11.43+(10-1)18.9}{10+8-2}=15.63$$

$$t=\frac{(\bar{x}_1-\bar{x}_2)-(\bar{u}_1-\bar{u}_2)}{\sqrt{s_p{}^2(\frac{1}{n_1}+\frac{1}{n_2})}}=\frac{(27.5-21.3)-(3)}{\sqrt{15.63(\frac{1}{8}+\frac{1}{10})}}=\frac{3.2}{1.86}=1.72$$

21. As 1.72 < 2.583 then accept H_0.

Using the T-TEST procedure and with the following SAS code

```
proc ttest data=eggs0 H0=3;

    class farm;

    var wtoz;

run;
```

the following output is generated:

The TTEST Procedure

Statistics

Variable	farm	N	Lower CL Mean	Mean	Upper CL Mean	Lower CL Std Dev	Std Dev	Upper CL Std Dev	Std Err
wtoz	1	8	23.317	27.5	31.683	1.9863	3.3806	8.9927	1.1952
wtoz	2	10	16.832	21.3	25.768	2.6853	4.3474	9.9017	1.3748
wtoz	Diff (1-2)		0.7224	6.2	11.678	2.7016	3.9536	6.974	1.8754

T-Tests

Variable	Method	Variances	DF	t Value	Pr > |t|
wtoz	Pooled	Equal	16	1.71	0.1073
wtoz	Satterthwaite	Unequal	16	1.76	0.0981

Equality of Variances

Variable	Method	Num DF	Den DF	F Value	Pr > F
wtoz	Folded F	9	7	1.65	0.5198

A decision is made to accept H_0 or H_1 by comparing the "Pr > |t|" value against the significance level – in this case, $0.005 < 0.1073$ so, accept H_0.

Distributions Tables

Statistical Distribution Tables

These include Standard Normal, Student t, Chi-Square, and Fisher F Distributions.

NOTES:

These are laid out in the traditional format of distribution tables as presented below and have the advantage of showing many values simultaneously; thus, they enable the user to examine and quickly explore ranges of probabilities.

The T, Chi, and F range of probabilities in these grids are limited to some of the popular α- values. For the F distribution, the α- values of probabilities are limited to α = 0.10 , 0.05 , 0.025, and 0.01 only.

Note that all table values were borrowed from https://documents.software.dell.com/statistics/

and were verified by them against other published tables.

APPENDIX IV (A): Standard Normal (Z) Table

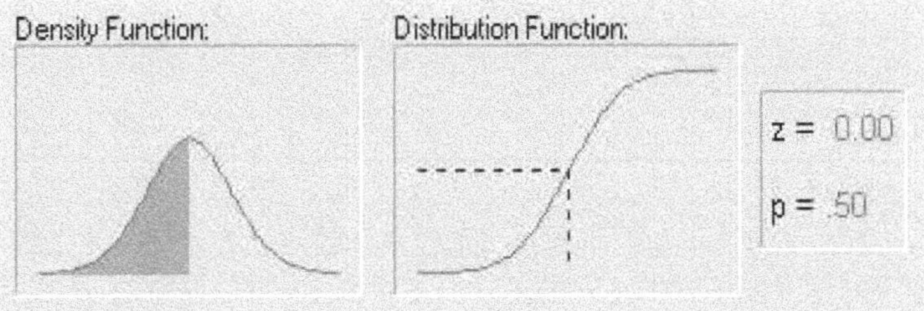

The Standard Normal distribution is used in various hypothesis tests, including tests on single means, the difference between two means, and tests on proportions. The Standard Normal distribution has a mean of 0 and a standard deviation of 1. The animation above shows various (left) tail areas for this distribution. For more information on the Normal Distribution as it is used in statistical testing, see Elementary Concepts. See also, Normal Distribution.

As shown in the illustration below, the values inside the given table represent the areas under the standard normal curve for values between 0 and the relative z-score. For example, to determine the area under the curve between 0 and 2.36, look in the intersecting cell for the row labeled 2.30 and the column labeled 0.06. The area under the curve is .4909. To determine the area between 0 and a negative value, look at the intersecting cell of the row and column, which sums to the absolute value of the number in question. For example, the area under the curve between -1.3 and 0 is equal to the area under the curve between 1.3 and 0, so look at the cell on the 1.3 rows and the 0.00

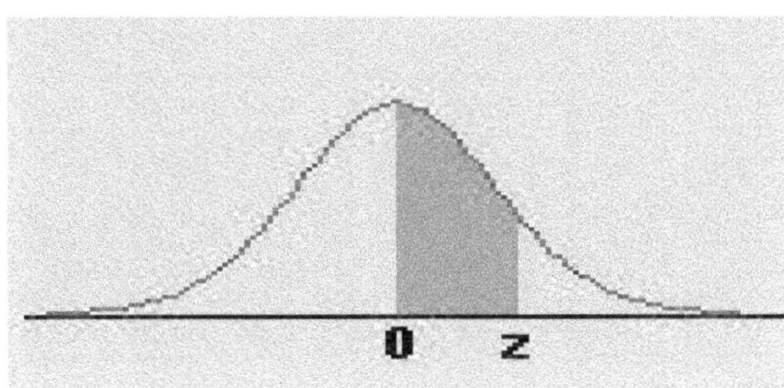

Column (the area is 0.4032).

The area between 0 and z

z	0.00	0.01	0.02	0.03	0.04	0.05	0.06	0.07	0.08	0.09
0.0	0.00	0.00	0.00	0.01	0.01	0.01	0.02	0.02	0.03	0.03
0.1	0.03	0.04	0.04	0.05	0.05	0.05	0.06	0.06	0.07	0.07
0.2	0.07	0.08	0.08	0.09	0.09	0.09	0.10	0.10	0.11	0.11
0.3	0.11	0.12	0.12	0.12	0.13	0.13	0.14	0.14	0.14	0.15
0.4	0.15	0.15	0.16	0.16	0.17	0.17	0.17	0.18	0.18	0.18
0.5	0.19	0.19	0.19	0.20	0.20	0.20	0.21	0.21	0.21	0.22
0.6	0.22	0.22	0.23	0.23	0.23	0.24	0.24	0.24	0.25	0.25
0.7	0.25	0.26	0.26	0.26	0.27	0.27	0.27	0.28	0.28	0.28
0.8	0.28	0.29	0.29	0.29	0.29	0.30	0.30	0.30	0.31	0.31
0.9	0.31	0.31	0.32	0.32	0.32	0.32	0.33	0.33	0.33	0.33
1.0	0.34	0.34	0.34	0.34	0.35	0.35	0.35	0.35	0.35	0.36
1.1	0.36	0.36	0.36	0.37	0.37	0.37	0.37	0.37	0.38	0.38
1.2	0.38	0.38	0.38	0.39	0.39	0.39	0.39	0.39	0.39	0.40
1.3	0.40	0.40	0.40	0.40	0.40	0.41	0.41	0.41	0.41	0.41
1.4	0.41	0.42	0.42	0.42	0.42	0.42	0.42	0.42	0.43	0.43
1.5	0.43	0.43	0.43	0.43	0.43	0.43	0.44	0.44	0.44	0.44
1.6	0.44	0.44	0.44	0.44	0.44	0.45	0.45	0.45	0.45	0.45
1.7	0.45	0.45	0.45	0.45	0.45	0.45	0.46	0.46	0.46	0.46
1.8	0.46	0.46	0.46	0.46	0.46	0.46	0.46	0.46	0.46	0.47
1.9	0.47	0.47	0.47	0.47	0.47	0.47	0.47	0.47	0.47	0.47
2.0	0.47	0.47	0.47	0.47	0.47	0.47	0.48	0.48	0.48	0.48
2.1	0.48	0.48	0.48	0.48	0.48	0.48	0.48	0.48	0.48	0.48
2.2	0.48	0.48	0.48	0.48	0.48	0.48	0.48	0.48	0.48	0.48
2.3	0.48	0.48	0.48	0.49	0.49	0.49	0.49	0.49	0.49	0.49
2.4	0.49	0.49	0.49	0.49	0.49	0.49	0.49	0.49	0.49	0.49
2.5	0.49	0.49	0.49	0.49	0.49	0.49	0.49	0.49	0.49	0.49
2.6	0.49	0.49	0.49	0.49	0.49	0.49	0.49	0.49	0.49	0.49
2.7	0.49	0.49	0.49	0.49	0.49	0.49	0.49	0.49	0.49	0.49
2.8	0.49	0.49	0.49	0.49	0.49	0.49	0.49	0.49	0.49	0.49
2.9	0.49	0.49	0.49	0.49	0.49	0.49	0.49	0.49	0.49	0.49
3.0	0.49	0.49	0.49	0.49	0.49	0.49	0.49	0.49	0.49	0.49

APPENDIX IV (B): Student's Table

The Shape of the Student's t distribution is determined by the degrees of freedom. As shown in the animation above, its shape changes as the degrees of freedom increases. For more information on how this distribution is used in hypothesis testing, see the t-test for independent samples and t-the est for dependent samples in Basic Statistics and Tables. See also, Student's t Distribution. As indicated by the chart below, the areas given at the top of this table are the right tail areas for the t-value inside the table. To determine the 0.05 critical value from the t-distribution with 6 degrees of freedom, look in the 0.05 column at the six rows$_{(.05,6)}$ = 1.943180.

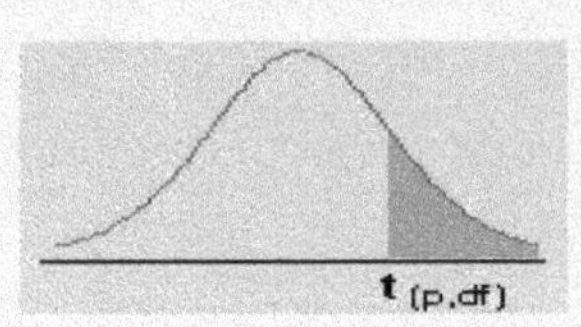

ttable with right tail probabilities

df \p	0.40	0.25	0.10	0.05	0.025	0.01	0.005	0.0005
1	0.324920	1.000000	3.077684	6.313752	12.70620	31.82052	63.65674	636.6192
2	0.288675	0.816497	1.885618	2.919986	4.30265	6.96456	9.92484	31.5991
3	0.276671	0.764892	1.637744	2.353363	3.18245	4.54070	5.84091	12.9240
4	0.270722	0.740697	1.533206	2.131847	2.77645	3.74695	4.60409	8.6103
5	0.267181	0.726687	1.475884	2.015048	2.57058	3.36493	4.03214	6.8688
6	0.264835	0.717558	1.439756	1.943180	2.44691	3.14267	3.70743	5.9588
7	0.263167	0.711142	1.414924	1.894579	2.36462	2.99795	3.49948	5.4079
8	0.261921	0.706387	1.396815	1.859548	2.30600	2.89646	3.35539	5.0413
9	0.260955	0.702722	1.383029	1.833113	2.26216	2.82144	3.24984	4.7809
10	0.260185	0.699812	1.372184	1.812461	2.22814	2.76377	3.16927	4.5869
11	0.259556	0.697445	1.363430	1.795885	2.20099	2.71808	3.10581	4.4370
12	0.259033	0.695483	1.356217	1.782288	2.17881	2.68100	3.05454	4.3178
13	0.258591	0.693829	1.350171	1.770933	2.16037	2.65031	3.01228	4.2208
14	0.258213	0.692417	1.345030	1.761310	2.14479	2.62449	2.97684	4.1405
15	0.257885	0.691197	1.340606	1.753050	2.13145	2.60248	2.94671	4.0728
16	0.257599	0.690132	1.336757	1.745884	2.11991	2.58349	2.92078	4.0150
17	0.257347	0.689195	1.333379	1.739607	2.10982	2.56693	2.89823	3.9651
18	0.257123	0.688364	1.330391	1.734064	2.10092	2.55238	2.87844	3.9216
19	0.256923	0.687621	1.327728	1.729133	2.09302	2.53948	2.86093	3.8834
20	0.256743	0.686954	1.325341	1.724718	2.08596	2.52798	2.84534	3.8495

21	0.256580	0.686352	1.323188	1.720743	2.07961	2.51765	2.83136	3.8193
22	0.256432	0.685805	1.321237	1.717144	2.07387	2.50832	2.81876	3.7921
23	0.256297	0.685306	1.319460	1.713872	2.06866	2.49987	2.80734	3.7676
24	0.256173	0.684850	1.317836	1.710882	2.06390	2.49216	2.79694	3.7454
25	0.256060	0.684430	1.316345	1.708141	2.05954	2.48511	2.78744	3.7251
26	0.255955	0.684043	1.314972	1.705618	2.05553	2.47863	2.77871	3.7066
27	0.255858	0.683685	1.313703	1.703288	2.05183	2.47266	2.77068	3.6896
28	0.255768	0.683353	1.312527	1.701131	2.04841	2.46714	2.76326	3.6739
29	0.255684	0.683044	1.311434	1.699127	2.04523	2.46202	2.75639	3.6594
30	0.255605	0.682756	1.310415	1.697261	2.04227	2.45726	2.75000	3.6460
inf	0.253347	0.674490	1.281552	1.644854	1.95996	2.32635	2.57583	3.2905

APPENDIX IV (C): Chi-Square Table

Like the Student's *t*-Distribution, the *Chi-square* distribution's shape is determined by its degrees of freedom. The animation above shows the shape of the *Chi-square* distribution as the degrees of freedom increase (1, 2, 5, 10, 25, and 50). For examples of tests of the hypothesis that use the *Chi-square distribution*, see Statistics in crosstabulation tables in Basic Statistics and Tables as well as Nonlinear Estimation. See also Chi-square Distribution. As shown in the illustration below, the values inside this table are critical values of the Chi-square distribution with the corresponding degrees of freedom. To determine the value from a Chi-square distribution (with a specific degree of freedom) which has a given area above it, go to the given area column and the desired degree of freedom row. For example, the .25 critical value for a Chi-square with 4 degrees of freedom is 5.38527. This means that the area to the right of 5.38527 in a Chi-square distribution with 4 degrees of freedom is .25.

Right   tail   areas   for   the *Chi-square* Distribution

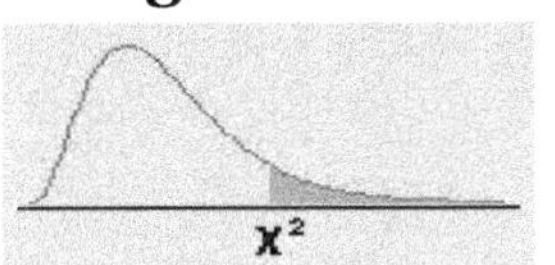

df\area	.995	.990	.975	.950	.900	.750	.500	.250	.100	.050	.025	.010	.005
1	0.00004	0.00016	0.00098	0.00393	0.01579	0.10153	0.45494	1.32330	2.70554	3.84146	5.02389	6.63490	7.87944
2	0.01003	0.02010	0.05064	0.10259	0.21072	0.57536	1.38629	2.77259	4.60517	5.99146	7.37776	9.21034	10.5966
3	0.07172	0.11483	0.21580	0.35185	0.58437	1.21253	2.36597	4.10834	6.25139	7.81473	9.34840	11.3448	12.8381
4	0.20699	0.29711	0.48442	0.71072	1.06362	1.92256	3.35669	5.38527	7.77944	9.48773	11.1432	13.2767	14.8602
5	0.41174	0.55430	0.83121	1.14548	1.61031	2.67460	4.35146	6.62568	9.23636	11.0705	12.8325	15.0862	16.7496
6	0.67573	0.87209	1.23734	1.63538	2.20413	3.45460	5.34812	7.84080	10.644	12.591	14.449	16.811	18.5475
7	0.98926	1.23904	1.68987	2.16735	2.83311	4.25485	6.34581	9.03715	12.017	14.067	16.012	18.475	20.277
8	1.34441	1.64650	2.17973	2.73264	3.48954	5.07064	7.34412	10.218	13.3615	15.5073	17.5345	20.0902	21.9549
9	1.73493	2.08790	2.70039	3.32511	4.16816	5.89883	8.34283	11.388	14.6836	16.918	19.022	21.665	23.589
10	2.15586	2.55821	3.24697	3.94030	4.86518	6.73720	9.34182	12.5488	15.9871	18.307	20.4831	23.2092	25.1881
11	2.60322	3.05348	3.81575	4.57481	5.57778	7.58414	10.34100	13.70069	17.27501	19.6751	21.9200	24.7249	26.7568
12	3.07382	3.57057	4.40379	5.22603	6.30380	8.43842	11.34032	14.84540	18.54935	21.02607	23.33666	26.21697	28.29952
13	3.56503	4.10692	5.00875	5.89186	7.04150	9.29907	12.33976	15.98391	19.81193	22.36203	24.73560	27.68825	29.81947
14	4.07467	4.66043	5.62873	6.57063	7.78953	10.16531	13.33927	17.11693	21.06414	23.68479	26.11895	29.14124	31.31935
15	4.60092	5.22935	6.26214	7.26094	8.54676	11.03654	14.33886	18.24509	22.30713	24.99579	27.48839	30.57791	32.80132
16	5.14221	5.81221	6.90766	7.96165	9.31224	11.91222	15.33850	19.36886	23.54183	26.29623	28.84535	31.99993	34.26719
17	5.69722	6.40776	7.56419	8.67176	10.08519	12.79193	16.33818	20.48868	24.76904	27.58711	30.19101	33.40866	35.71847
18	6.26480	7.01491	8.23075	9.39046	10.86494	13.67529	17.33790	21.60489	25.98942	28.86930	31.52638	34.80531	37.15645
19	6.84397	7.63273	8.90652	10.11701	11.65091	14.56200	18.33765	22.71781	27.20357	30.14353	32.85233	36.19087	38.58226
20	7.43384	8.26040	9.59078	10.85081	12.44261	15.45177	19.33743	23.82769	28.41198	31.41043	34.16961	37.56623	39.99685
21	8.03365	8.89720	10.28290	11.59131	13.23960	16.34438	20.33723	24.93478	29.61509	32.67057	35.47888	38.93217	41.40106
22	8.64272	9.54249	10.98232	12.33801	14.04149	17.23962	21.33704	26.03927	30.81328	33.92444	36.78071	40.28936	42.79565
23	9.26042	10.19572	11.68855	13.09051	14.84796	18.13730	22.33688	27.14134	32.00690	35.17246	38.07563	41.63840	44.18128
24	9.88623	10.85636	12.40115	13.84843	15.65868	19.03725	23.33673	28.24115	33.19624	36.41503	39.36408	42.97982	45.55851
25	10.51965	11.52398	13.11972	14.61141	16.47341	19.93934	24.33659	29.33885	34.38159	37.65248	40.64647	44.31410	46.92789
26	11.16024	12.19815	13.84390	15.37916	17.29188	20.84343	25.33646	30.43457	35.56317	38.88514	41.92317	45.64168	48.28988
27	11.80759	12.87850	14.57338	16.15140	18.11390	21.74940	26.33634	31.52841	36.74122	40.11327	43.19451	46.96294	49.64492
28	12.46134	13.56471	15.30786	16.92788	18.93924	22.65716	27.33623	32.62049	37.91592	41.33714	44.46079	48.27824	50.99338
29	13.12115	14.25645	16.04707	17.70837	19.76774	23.56659	28.33613	33.71091	39.08747	42.55697	45.72229	49.58788	52.33562
30	13.78672	14.95346	16.79077	18.49266	20.59923	24.47761	29.33603	34.79974	40.25602	43.77297	46.97924	50.89218	53.67196

APPENDIX IV (D): F Distribution Tables

The F distribution is a right-skewed distribution used most commonly in the Analysis of Variance (see ANOVA). The F distribution is a ratio of two *Chi-square* distributions, and a specific F distribution is denoted by the degrees of freedom for the numerator Chi-square and the degrees of freedom for the denominator Chi-square. An example of the $F_{(10,10)}$ distribution is shown in the animation above. When referencing the F distribution, the numerator degrees of freedom is always given first, as switching the order of degrees of freedom changes the distribution (e.g., $F_{(10,12)}$ does not equal $F_{(12,10)}$). For the four F tables below, the rows represent denominator degrees of freedom, and the columns represent numerator degrees of freedom. The right tail area is given in the name of the table. For example, to determine the .05 critical value for an F distribution with 10 and 12 degrees of freedom, look in the ten columns (numerator) and 12 rows (denominator) of the F Table for alpha=.05. $F_{(.05, 10, 12)} = 2.7534$.

In the following pages, we have the different tables for the four different **α- values of 0.10, 0.05, 02,5, and 0.01; each laid out on a different page.**

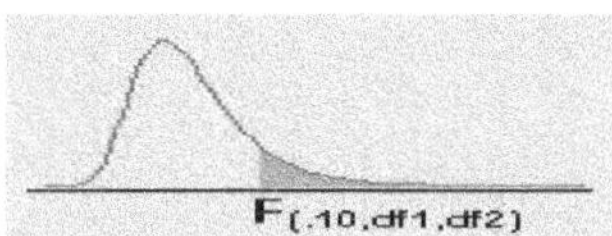

df2 df1	1	2	3	4	5	6	7	8	9	10	12	15	20	24	30
1	39.8	49.	53.5	55.8	57.2	58.2	58.9	59.4	59.8	60.1	60.7	61.2	61.7	62.0	62.2
2	8.52	9.0	9.16	9.24	9.29	9.32	9.34	9.36	9.38	9.39	9.40	9.42	9.44	9.44	9.45
3	5.53	5.4	5.39	5.34	5.30	5.28	5.26	5.25	5.24	5.23	5.21	5.20	5.18	5.17	5.16
4	4.54	4.3	4.19	4.10	4.05	4.00	3.97	3.95	3.93	3.91	3.89	3.87	3.84	3.83	3.81
5	4.0	3.7	3.61	3.52	3.45	3.40	3.36	3.33	3.31	3.29	3.26	3.23	3.20	3.19	3.17
6	3.77	3.4	3.28	3.18	3.10	3.05	3.01	2.98	2.95	2.93	2.90	2.87	2.83	2.81	2.79
7	3.58	3.2	3.07	2.96	2.88	2.82	2.78	2.75	2.72	2.70	2.66	2.63	2.59	2.57	2.55
8	3.45	3.11	2.92	2.80	2.72	2.66	2.62	2.58	2.56	2.53	2.50	2.46	2.42	2.40	2.38
9	3.36	3.0	2.81	2.69	2.61	2.55	2.50	2.46	2.44	2.41	2.37	2.33	2.29	2.27	2.25
10	3.28	2.9	2.72	2.60	2.52	2.46	2.41	2.37	2.34	2.32	2.28	2.24	2.20	2.17	2.15
11	3.22	2.8	2.66	2.53	2.45	2.38	2.34	2.30	2.27	2.24	2.20	2.16	2.12	2.10	2.07
12	3.17	2.8	2.60	2.48	2.39	2.33	2.28	2.24	2.21	2.18	2.14	2.10	2.05	2.03	2.01
13	3.13	2.7	2.56	2.43	2.34	2.28	2.23	2.19	2.16	2.13	2.09	2.05	2.00	1.98	1.95
14	3.10	2.7	2.52	2.39	2.30	2.24	2.19	2.15	2.12	2.09	2.05	2.00	1.96	1.93	1.911
15	3.07	2.6	2.48	2.36	2.27	2.20	2.15	2.11	2.08	2.05	2.01	1.97	1.92	1.89	1.87
16	3.0	2.6	2.46	2.33	2.24	2.17	2.12	2.08	2.05	2.02	1.98	1.93	1.89	1.86	1.83
17	3.02	2.6	2.43	2.30	2.21	2.15	2.10	2.06	2.02	2.00	1.95	1.911	1.86	1.83	1.80
18	3.0	2.6	2.41	2.28	2.19	2.12	2.07	2.03	2.00	1.97	1.93	1.88	1.83	1.81	1.78
19	2.98	2.6	2.39	2.26	2.17	2.10	2.05	2.01	1.98	1.95	1.911	1.86	1.81	1.78	1.75
20	2.97	2.5	2.38	2.24	2.15	2.09	2.03	1.99	1.96	1.93	1.89	1.84	1.79	1.76	1.73
21	2.96	2.5	2.36	2.23	2.14	2.07	2.02	1.98	1.94	1.91	1.87	1.82	1.77	1.74	1.71
22	2.94	2.5	2.35	2.21	2.12	2.06	2.00	1.96	1.93	1.90	1.85	1.811	1.75	1.731	1.70
23	2.93	2.5	2.33	2.20	2.11	2.04	1.99	1.95	1.91	1.89	1.84	1.79	1.74	1.715	1.68
24	2.92	2.5	2.32	2.19	2.10	2.03	1.98	1.94	1.90	1.87	1.83	1.78	1.73	1.701	1.67
25	2.917	2.5	2.31	2.18	2.09	2.02	1.971	1.92	1.89	1.86	1.82	1.77	1.717	1.68	1.65
26	2.9	2.5	2.30	2.17	2.08	2.01	1.96	1.91	1.88	1.85	1.80	1.75	1.70	1.67	1.64
27	2.9	2.5	2.29	2.16	2.07	2.00	1.95	1.90	1.87	1.84	1.79	1.74	1.69	1.66	1.63
28	2.89	2.5	2.29	2.15	2.06	1.99	1.94	1.90	1.86	1.83	1.78	1.73	1.68	1.65	1.62
29	2.8	2.4	2.28	2.14	2.05	1.98	1.93	1.89	1.85	1.82	1.78	1.73	1.67	1.64	1.61
30	2.8	2.4	2.27	2.14	2.04	1.98	1.92	1.88	1.84	1.81	1.77	1.72	1.66	1.63	1.60
40	2.83	2.4	2.22	2.09	1.99	1.92	1.87	1.82	1.79	1.76	1.71	1.66	1.60	1.57	1.54

continued

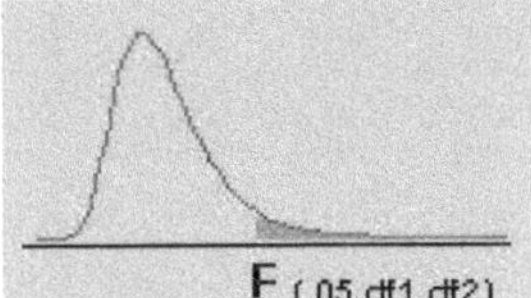

df2/df1	11	22	33	44	55	66	77	88	99	110	112	115	220	224	330	440	60	1120	I∞
11	161.4	199.	215.	224.	230.	233.	236.	238.	240.	241.	243.	245.	248.	249.	250.	251.	252.	253.	254.
22	18.51	19.0	19.1	19.2	19.2	19.3	19.3	19.3	19.3	19.3	19.4	19.4	19.4	19.4	19.4	19.4	19.4	19.4	19.4
33	10.12	9.55	9.27	9.11	9.01	8.94	8.88	8.84	8.81	8.78	8.74	8.70	8.66	8.63	8.61	8.59	8.57	8.54	8.52
4	7.708	6.94	6.59	6.38	6.25	6.16	6.09	6.04	5.99	5.96	5.91	5.85	5.80	5.77	5.74	5.71	5.68	5.65	5.62
5	6.607	5.78	5.40	5.19	5.05	4.95	4.87	4.81	4.77	4.73	4.67	4.61	4.55	4.52	4.49	4.46	4.43	4.39	4.36
6	5.987	5.14	4.75	4.53	4.38	4.28	4.20	4.14	4.09	4.06	3.99	3.93	3.87	3.84	3.80	3.77	3.73	3.70	3.66
7	5.591	4.73	4.34	4.12	3.97	3.86	3.78	3.72	3.67	3.63	3.57	3.51	3.44	3.41	3.37	3.34	3.30	3.26	3.22
8	5.317	4.45	4.06	3.83	3.68	3.58	3.50	3.43	3.38	3.34	3.28	3.21	3.15	3.11	3.07	3.04	3.00	2.96	2.92
9	5.1174	4.25	3.86	3.63	3.48	3.37	3.29	3.22	3.17	3.13	3.07	3.00	2.93	2.90	2.86	2.82	2.78	2.74	2.70
10	4.964	4.10	3.70	3.47	3.32	3.21	3.13	3.07	3.02	2.97	2.91	2.84	2.77	2.73	2.69	2.66	2.62	2.58	2.53
11	4.844	3.98	3.58	3.35	3.20	3.09	3.01	2.94	2.89	2.85	2.78	2.71	2.64	2.60	2.57	2.53	2.49	2.44	2.40
12	4.747	3.88	3.49	3.25	3.10	2.99	2.91	2.84	2.79	2.75	2.68	2.61	2.54	2.50	2.46	2.42	2.38	2.34	2.29
13	4.667	3.80	3.41	3.17	3.02	2.91	2.83	2.76	2.71	2.67	2.60	2.53	2.45	2.42	2.38	2.33	2.29	2.25	2.20
14	4.600	3.73	3.34	3.11	2.95	2.84	2.76	2.69	2.64	2.60	2.53	2.46	2.38	2.34	2.30	2.26	2.22	2.17	2.13
15	4.543	3.68	3.28	3.05	2.90	2.79	2.70	2.64	2.58	2.54	2.47	2.40	2.32	2.28	2.24	2.20	2.16	2.11	2.06
16	4.494	3.63	3.23	3.00	2.85	2.74	2.65	2.59	2.53	2.49	2.42	2.35	2.27	2.23	2.19	2.15	2.10	2.05	2.00
17	4.451	3.59	3.19	2.96	2.81	2.69	2.61	2.54	2.49	2.44	2.38	2.30	2.23	2.18	2.14	2.10	2.05	2.01	1.96
18	4.413	3.55	3.15	2.92	2.77	2.66	2.57	2.51	2.45	2.41	2.34	2.26	2.19	2.14	2.10	2.06	2.01	1.96	1.91
19	4.380	3.52	3.12	2.89	2.74	2.62	2.54	2.47	2.42	2.37	2.30	2.23	2.15	2.11	2.07	2.02	1.97	1.93	1.87
20	4.351	3.49	3.09	2.86	2.71	2.59	2.51	2.44	2.39	2.34	2.27	2.20	2.12	2.08	2.03	1.99	1.94	1.89	1.84
21	4.324	3.46	3.07	2.84	2.68	2.57	2.48	2.42	2.36	2.32	2.25	2.17	2.09	2.05	2.01	1.96	1.91	1.86	1.81
22	4.300	3.44	3.04	2.81	2.66	2.54	2.46	2.39	2.34	2.29	2.22	2.15	2.07	2.02	1.98	1.93	1.88	1.83	1.78
23	4.279	3.42	3.02	2.79	2.64	2.52	2.44	2.37	2.32	2.27	2.20	2.12	2.04	2.00	1.96	1.91	1.86	1.81	1.75
24	4.259	3.40	3.00	2.73	2.62	2.50	2.42	2.35	2.30	2.25	2.18	2.10	2.02	1.98	1.93	1.89	1.84	1.78	1.73
25	4.241	3.38	2.99	2.75	2.60	2.49	2.40	2.33	2.28	2.23	2.16	2.08	2.00	1.96	1.91	1.87	1.82	1.76	1.711
26	4.225	3.36	2.97	2.74	2.58	2.47	2.38	2.32	2.26	2.21	2.14	2.07	1.98	1.94	1.90	1.85	1.80	1.74	1.69
27	4.210	3.35	2.96	2.72	2.57	2.45	2.37	2.30	2.25	2.20	2.13	2.05	1.97	1.92	1.88	1.83	1.78	1.73	1.67
28	4.196	3.34	2.94	2.71	2.55	2.44	2.35	2.29	2.23	2.19	2.117	2.04	1.95	1.91	1.86	1.82	1.76	1.713	1.65
29	4.183	3.32	2.93	2.70	2.54	2.43	2.34	2.27	2.22	2.17	2.10	2.02	1.94	1.90	1.85	1.80	1.75	1.69	1.63
30	4.170	3.31	2.92	2.68	2.53	2.42	2.33	2.26	2.21	2.16	2.09	2.01	1.93	1.88	1.84	1.79	1.73	1.68	1.62
40	4.084	3.23	2.83	2.60	2.44	2.33	2.24	2.18	2.12	2.07	2.00	1.92	1.83	1.79	1.74	1.69	1.63	1.57	1.50
60	4.001	3.15	2.75	2.52	2.36	2.25	2.16	2.09	2.04	1.99	1.91	1.83	1.74	1.70	1.64	1.59	1.53	1.46	1.38
120	3.920	3.07	2.68	2.44	2.28	2.17	2.08	2.01	1.95	1.9	1.83	1.75	1.65	1.60	1.55	1.49	1.42	1.35	1.25
∞	3.841	2.99	2.60	2.37	2.21	2.09	2.00	1.93	1.87	1.83	1.75	1.66	1.57	1.517	1.45	1.34	1.31	1.22	1.00

continued

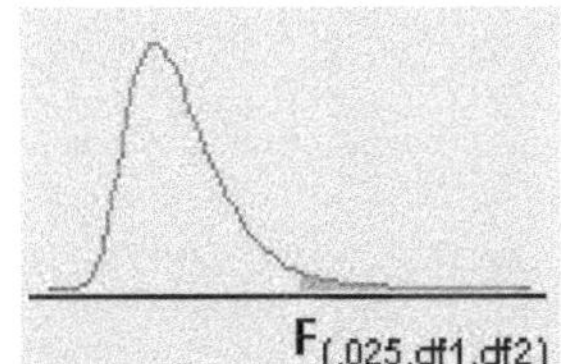

df2/df1	11	22	33	44	55	66	77	88	99	110	112	115	220	224	330	440	660	1120	INF
1	647.7	799.	864.	899.	921.	937.	948.	956.	963.	968.	976.	984.	993.	997.	1001	1005	1009	1014	1018
2	38.5	39.0	39.1	39.2	39.2	39.3	39.3	39.3	39.3	39.3	39.4	39.4	39.4	39.4	39.4	39.4	39.4	39.4	39.4
3	17.44	16.0	15.4	15.1	14.8	14.7	14.6	14.5	14.4	14.4	14.3	14.2	14.1	14.1	14.0	14.0	13.9	13.9	13.9
4	12.21	10.6	9.97	9.60	9.36	9.19	9.07	8.97	8.90	8.84	8.75	8.65	8.55	8.51	8.46	8.41	8.36	8.30	8.25
5	10.0	8.43	7.76	7.38	7.14	6.97	6.85	6.75	6.68	6.61	6.52	6.42	6.32	6.27	6.22	6.17	6.12	6.06	6.01
6	8.813	7.25	6.59	6.22	5.98	5.81	5.69	5.59	5.52	5.46	5.36	5.26	5.16	5.117	5.06	5.01	4.95	4.90	4.84
7	8.072	6.54	5.88	5.52	5.28	5.11	4.99	4.89	4.82	4.76	4.66	4.56	4.46	4.41	4.36	4.30	4.25	4.19	4.14
8	7.570	6.05	5.41	5.05	4.81	4.65	4.52	4.43	4.35	4.29	4.19	4.10	3.99	3.94	3.89	3.84	3.78	3.72	3.67
9	7.209	5.71	5.07	4.711	4.48	4.31	4.19	4.10	4.02	3.96	3.86	3.76	3.66	3.61	3.56	3.50	3.44	3.39	3.33
110	6.936	5.45	4.82	4.46	4.23	4.07	3.94	3.85	3.77	3.71	3.62	3.52	3.41	3.36	3.311	3.25	3.19	3.14	3.08
111	6.724	5.25	4.63	4.27	4.04	3.88	3.75	3.66	3.58	3.52	3.42	3.32	3.22	3.17	3.11	3.06	3.00	2.94	2.88
112	6.553	5.09	4.47	4.12	3.89	3.72	3.60	3.511	3.43	3.37	3.27	3.177	3.07	3.01	2.96	2.90	2.84	2.78	2.72
113	6.41	4.96	4.34	3.99	3.77	3.60	3.48	3.38	3.31	3.24	3.15	3.05	2.94	2.89	2.83	2.78	2.72	2.65	2.59
114	6.297	4.85	4.24	3.89	3.66	3.50	3.37	3.28	3.20	3.14	3.05	2.94	2.84	2.78	2.73	2.67	2.61	2.55	2.48
115	6.199	4.76	4.15	3.80	3.57	3.41	3.29	3.19	3.12	3.06	2.96	2.86	2.75	2.70	2.64	2.58	2.52	2.46	2.39
116	6.115	4.68	4.07	3.72	3.50	3.34	3.21	3.12	3.04	2.98	2.88	2.78	2.68	2.62	2.56	2.50	2.44	2.38	2.31
117	6.04	4.61	4.01	3.66	3.43	3.27	3.15	3.06	2.98	2.92	2.82	2.72	2.61	2.55	2.50	2.44	2.38	2.31	2.24
118	5.978	4.56	3.95	3.60	3.38	3.22	3.09	3.00	2.92	2.86	2.76	2.66	2.55	2.50	2.44	2.38	2.32	2.25	2.18
119	5.921	4.50	3.90	3.55	3.33	3.171	3.05	2.95	2.88	2.81	2.71	2.61	2.50	2.45	2.39	2.33	2.27	2.20	2.13
220	5.871	4.46	3.85	3.51	3.28	3.12	3.00	2.91	2.83	2.77	2.67	2.57	2.46	2.40	2.34	2.28	2.22	2.15	2.08
21	5.826	4.41	3.81	3.47	3.25	3.08	2.96	2.87	2.79	2.73	2.63	2.53	2.42	2.36	2.30	2.24	2.18	2.11	2.04
22	5.786	4.38	3.78	3.44	3.21	3.05	2.93	2.83	2.76	2.69	2.60	2.49	2.38	2.33	2.27	2.21	2.14	2.07	2.00
23	5.749	4.34	3.75	3.40	3.18	3.02	2.90	2.80	2.73	2.66	2.56	2.46	2.35	2.29	2.23	2.17	2.111	2.04	1.96
24	5.716	4.31	3.72	3.37	3.15	2.99	2.87	2.77	2.70	2.63	2.54	2.43	2.32	2.26	2.20	2.14	2.08	2.01	1.93
25	5.686	4.29	3.69	3.35	3.12	2.96	2.84	2.75	2.67	2.61	2.51	2.41	2.30	2.24	2.18	2.11	2.05	1.98	1.90
26	5.658	4.26	3.66	3.32	3.10	2.94	2.82	2.72	2.65	2.58	2.49	2.38	2.27	2.21	2.15	2.09	2.02	1.95	1.87
27	5.633	4.24	3.64	3.30	3.08	2.92	2.80	2.70	2.63	2.56	2.46	2.36	2.25	2.19	2.13	2.06	2.00	1.93	1.85
28	5.60	4.22	3.62	3.28	3.06	2.90	2.78	2.68	2.61	2.54	2.44	2.34	2.23	2.17	2.112	2.04	1.98	1.90	1.82
29	5.587	4.20	3.60	3.26	3.04	2.88	2.76	2.66	2.59	2.52	2.42	2.32	2.21	2.15	2.09	2.02	1.95	1.88	1.80
330	5.567	4.18	3.58	3.24	3.02	2.86	2.74	2.65	2.57	2.511	2.41	2.30	2.19	2.13	2.07	2.00	1.94	1.86	1.78
440	5.423	4.05	3.46	3.12	2.90	2.74	2.62	2.52	2.45	2.38	2.28	2.18	2.06	2.00	1.94	1.87	1.80	1.72	1.63
660	5.285	3.92	3.34	3.00	2.78	2.62	2.50	2.41	2.33	2.27	2.16	2.06	1.94	1.88	1.81	1.74	1.66	1.58	1.48
1120	5.15	3.8	3.2	2.8	2.67	2.51	2.39	2.29	2.22	2.15	2.05	1.94	1.82	1.75	1.69	1.61	1.53	1.43	1.31
inf	5.023	3.68	3.11	2.78	2.56	2.40	2.28	2.19	2.113	2.04	1.94	1.83	1.70	1.64	1.56	1.48	1.38	1.26	1.00

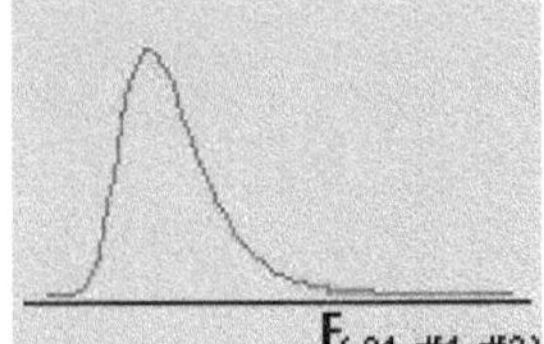

df2/df1	1	2	3	4	5	6	7	8	9	10	12	15	20	24	30	40	60	120	INF
1	405	4999.	5403.	5624.	5763.	5858.	5928.	5981.	6022.	6055.	6106.	6157.	6208.	6234.	6260.	6286.	6313.	6339.	6365.
2	98.5	99.00	99.16	99.24	99.29	99.33	99.35	99.37	99.38	99.39	99.41	99.43	99.44	99.45	99.46	99.47	99.48	99.49	99.49
3	34.116	30.817	29.457	28.710	28.237	27.911	27.672	27.489	27.345	27.229	27.052	26.872	26.690	26.598	26.505	26.411	26.316	26.221	26.125
4	21.1	18.00	16.69	15.97	15.52	15.20	14.97	14.79	14.65	14.54	14.37	14.19	14.02	13.92	13.83	13.74	13.65	13.55	13.46
5	16.2	13.27	12.06	11.39	10.96	10.67	10.45	10.28	10.15	10.0	9.88	9.722	9.553	9.466	9.379	9.291	9.202	9.112	9.020
6	13.7	10.92	9.780	9.148	8.746	8.466	8.260	8.102	7.976	7.874	7.718	7.559	7.396	7.313	7.229	7.143	7.057	6.969	6.880
7	12.2	9.547	8.451	7.847	7.460	7.191	6.993	6.840	6.719	6.620	6.469	6.314	6.155	6.074	5.992	5.908	5.824	5.737	5.650
8	11.2	8.649	7.591	7.006	6.632	6.371	6.178	6.029	5.911	5.814	5.667	5.515	5.359	5.279	5.198	5.116	5.032	4.946	4.859
9	10.5	8.022	6.992	6.422	6.057	5.802	5.613	5.467	5.351	5.257	5.111	4.962	4.808	4.729	4.649	4.567	4.483	4.398	4.311
10	10.0	7.559	6.552	5.994	5.636	5.386	5.200	5.057	4.942	4.849	4.706	4.558	4.405	4.327	4.247	4.165	4.082	3.996	3.909
11	9.64	7.206	6.217	5.668	5.316	5.069	4.886	4.744	4.632	4.539	4.397	4.251	4.099	4.021	3.941	3.860	3.776	3.690	3.602
12	9.33	6.927	5.953	5.412	5.064	4.821	4.640	4.499	4.388	4.296	4.155	4.010	3.858	3.780	3.701	3.619	3.535	3.449	3.361
13	9.07	6.701	5.739	5.205	4.862	4.620	4.441	4.302	4.191	4.100	3.960	3.815	3.665	3.587	3.507	3.425	3.341	3.255	3.165
14	8.86	6.515	5.564	5.035	4.695	4.456	4.278	4.140	4.030	3.939	3.80	3.656	3.505	3.427	3.348	3.266	3.181	3.094	3.004
15	8.68	6.359	5.417	4.893	4.556	4.318	4.142	4.004	3.895	3.805	3.666	3.522	3.372	3.294	3.214	3.132	3.047	2.959	2.868
16	8.53	6.226	5.292	4.773	4.437	4.202	4.026	3.890	3.780	3.691	3.553	3.409	3.259	3.181	3.101	3.018	2.933	2.845	2.753
17	8.40	6.112	5.185	4.669	4.336	4.102	3.927	3.791	3.682	3.593	3.455	3.312	3.162	3.084	3.003	2.920	2.835	2.746	2.653
18	8.28	6.013	5.092	4.579	4.248	4.015	3.841	3.705	3.597	3.508	3.371	3.227	3.077	2.999	2.919	2.835	2.749	2.660	2.566
19	8.18	5.926	5.010	4.500	4.171	3.939	3.765	3.631	3.523	3.434	3.297	3.153	3.003	2.925	2.844	2.761	2.674	2.584	2.489
20	8.09	5.849	4.938	4.431	4.103	3.871	3.699	3.564	3.457	3.368	3.231	3.088	2.938	2.859	2.778	2.695	2.608	2.517	2.421
21	8.01	5.780	4.874	4.369	4.042	3.812	3.640	3.506	3.398	3.310	3.173	3.03	2.880	2.801	2.720	2.636	2.548	2.457	2.360
22	7.94	5.719	4.817	4.313	3.988	3.758	3.587	3.453	3.346	3.258	3.121	2.978	2.827	2.749	2.667	2.583	2.495	2.403	2.305
23	7.88	5.664	4.765	4.264	3.939	3.710	3.539	3.406	3.299	3.211	3.074	2.931	2.781	2.702	2.620	2.535	2.447	2.354	2.256
24	7.82	5.614	4.718	4.218	3.895	3.667	3.496	3.363	3.256	3.168	3.032	2.88	2.738	2.659	2.577	2.492	2.403	2.310	2.211
25	7.77	5.568	4.675	4.177	3.855	3.627	3.457	3.324	3.217	3.129	2.993	2.850	2.699	2.620	2.538	2.453	2.364	2.270	2.169

26	7.72	5.526	4.637	4.140	3.818	3.591	3.421	3.288	3.182	3.094	2.958	2.815	2.664	2.585	2.503	2.417	2.327	2.233	2.131
27	7.67	5.488	4.601	4.106	3.785	3.558	3.388	3.256	3.149	3.062	2.926	2.783	2.632	2.552	2.470	2.384	2.294	2.198	2.097
28	7.63	5.453	4.568	4.074	3.754	3.528	3.358	3.226	3.120	3.032	2.896	2.753	2.602	2.522	2.440	2.354	2.263	2.167	2.064
29	7.59	5.420	4.538	4.045	3.725	3.499	3.330	3.198	3.092	3.005	2.86	2.726	2.574	2.495	2.412	2.325	2.234	2.138	2.034
30	7.56	5.39	4.51	4.01	3.6	3.47	3.3	3.17	3.067	2.97	2.843	2.700	2.549	2.469	2.386	2.299	2.208	2.111	2.006
40	7.31	5.17	4.31	3.8	3.51	3.29	3.12	2.9	2.8	2.8	2.6	2.5	2.36	2.2	2.2	2.11	2.01	1.91	1.80
60	7.0	4.97	4.12	3.6	3.33	3.11	2.95	2.8	2.71	2.63	2.4	2.3	2.19	2.11	2.0	1.93	1.83	1.72	1.60
120	6.8	4.78	3.9	3.4	3.17	2.95	2.79	2.6	2.55	2.47	2.3	2.19	2.0	1.95	1.86	1.76	1.65	1.53	1.38
inf	6.6	4.6	3.78	3.31	3.01	2.8	2.63	2.51	2.4	2.32	2.18	2.0	1.87	1.79	1.69	1.59	1.47	1.32	1.00

APPENDIX V: SAS Output Case

Obs	plant	rep	ht	npod_pl	Nplant in_m	nseeds_pl	TotBiomas	YIELD_kg_ha	daysMATURITY
1	Mungbean	1	0.5	26	24	9	10980	4126.7	90
2	Mungbean	2	0.4	22	33	8	9660	3767.3	90
3	Mungbean	3	0.5	25	30	7	4170	6477.3	90
4	Mungbean	4	0.5	15	32	9	7170	5604.2	90
5	Mungbean	1	1.1	53	22	9	30750	5902.4	95
6	Mungbean	2	1.1	25	18	8	31030	2024.8	95
7	Mungbean	3	1.1	50	28	7	26940	6001.0	95
8	Mungbean	4	1.1	26	27	9	30660	3782.8	95
9	Mungbean	1	0.6	17	30	12	22740	4136.2	90
10	Mungbean	2	0.5	34	27	12	30920	7045.4	90
11	Mungbean	3	0.6	31	29	9	20420	5468.3	90
12	Mungbean	4	0.5	22	23	14	33100	4594.9	90

Obs	plant	rep	ht	npod_pl	Nplant in_m	nseeds_pl	TotBiomas	YIELD_kg_ha	daysMATURITY
13	Mungbean	1	0.7	15	27	14	30480	3214.8	100
14	Mungbean	2	0.4	43	15	11	20260	4151.5	100
15	Mungbean	3	0.6	28	22	12	28860	4124.1	100
16	Mungbean	4	0.5	22	27	11	31690	3734.4	100
17	PigeonPe	1	0.9	0	11	0	830	0.0	120
18	PigeonPe	2	0.9	2	10	3	940	98.0	120
19	PigeonPe	3	0.9	10	8	3	1110	680.4	120
20	PigeonPe	4	0.9	21	10	2	1070	3429.2	120
21	PigeonPe	1	1.0	26	14	5	1520	143.1	130
22	PigeonPe	2	1.1	32	7	5	6160	901.8	130
23	PigeonPe	3	1.2	6	9	4	24070	2864.9	130
24	PigeonPe	4	1.2	8	7	5	21690	4225.8	130
25	PigeonPe	1	0.9	31	17	6	7090	3729.1	145
26	PigeonPe	2	0.9	27	16	4	8920	2037.9	145

Obs	plant	rep	ht	npod_pl	Nplant in_m	nseeds_pl	TotBiomas	YIELD_kg_ha	daysMATURITY
27	PigeonPe	3	0.8	26	10	4	8690	2903.0	145
28	PigeonPe	4	1.1	5	18	4	20100	429.7	145
29	PigeonPe	1	1.0	33	11	3	10070	1058.5	150
30	PigeonPe	2	1.0	43	11	5	19250	2258.4	150
31	PigeonPe	3	1.0	29	10	3	9070	789.1	150
32	PigeonPe	4	0.9	38	19	5	9410	3559.7	150

The SAS System

The REG Procedure

Model: eq1

Dependent Variable: YIELD_kg_ha

plant=Mungbean

Number of Observations Read 16

Number of Observations Used 16

Analysis of Variance

Source	DF	Sum of Squares	Mean Square	F Value	Pr > F
Model	1	4942959	4942959	3.22	0.0944
Error	14	21491044	1535075		
Corrected Total	15	26434003			

Root MSE	1238.98126	R-Square	0.1870
Dependent Mean	4634.75625	Adj R-Sq	0.1289
Coeff Var	26.73239		

Parameter Estimates

Variable	DF	Parameter Estimate	Standard Error	t Value	Pr > |t|
Intercept	1	3213.33171	850.53397	3.78	0.0020
npod_pl	1	50.09426	27.91639	1.79	0.0944

The SAS System

The REG Procedure

Model: eq1

Dependent Variable: YIELD_kg_ha

plant=Mungbean

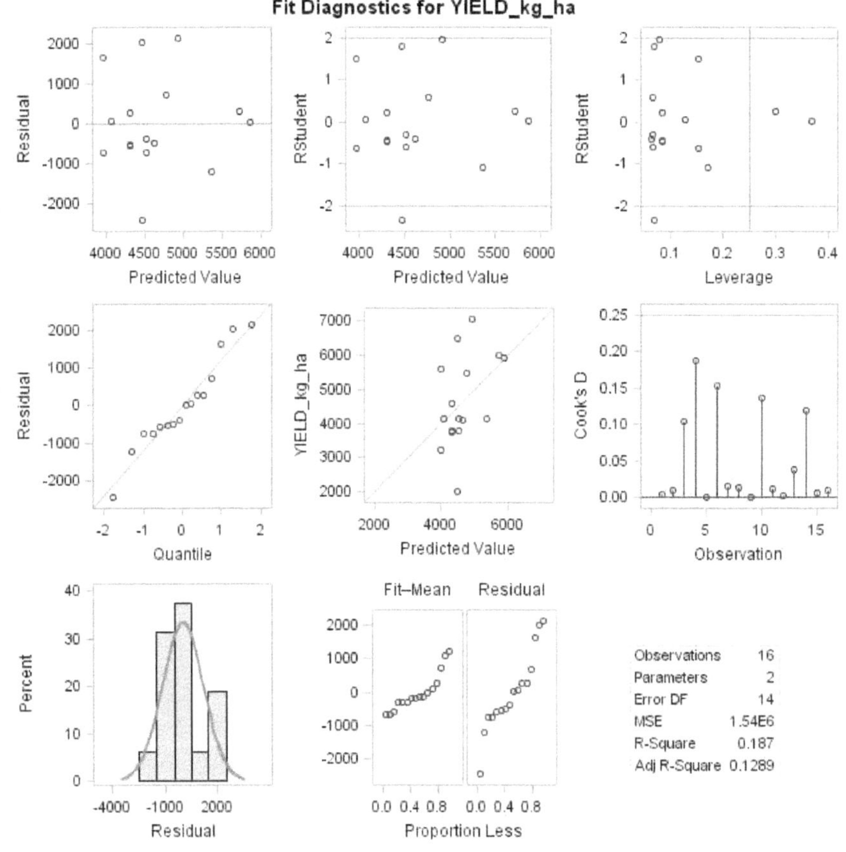

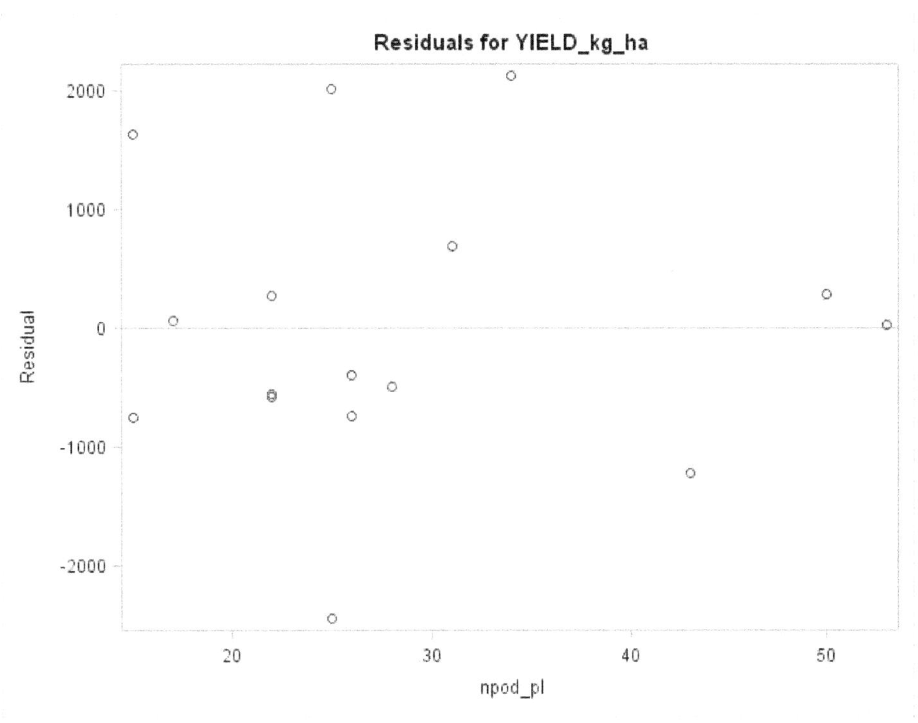

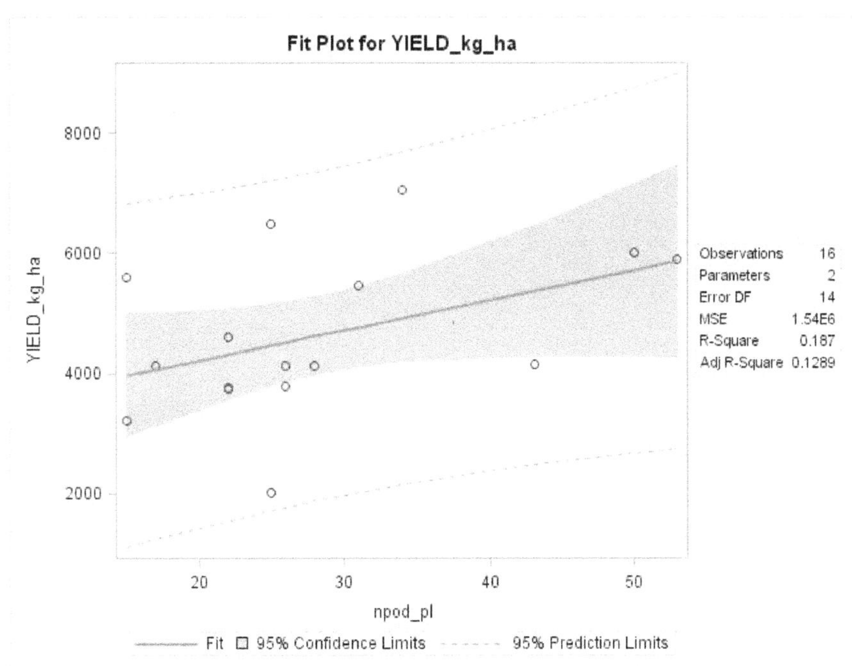

191

The REG Procedure

Model: eq2

Dependent Variable: YIELD_kg_ha

plant=Mungbean

Number of Observations Read	16
Number of Observations Used	16

Analysis of Variance

Source	DF	Sum of Squares	Mean Square	F Value	Pr > F
Model	2	8272040	4136020	2.96	0.0872
Error	13	18161963	1397074		
Corrected Total	15	26434003			

Root MSE	1181.97887	**R-Square** 0.3129
Dependent Mean	4634.75625	**Adj R-Sq** 0.2072
Coeff Var	25.50250	

Parameter Estimates

Variable	DF	Parameter Estimate	Standard Error	t Value	Pr > |t|
Intercept	1	4063.64634	980.71503	4.14	0.0012
npod_pl	1	59.98763	27.39234	2.19	0.0474
TotBiomas	1	-0.04893	0.03170	-1.54	0.1467

The SAS System

The REG Procedure

Model: eq2

Dependent Variable: YIELD_kg_ha

plant=Mungbean

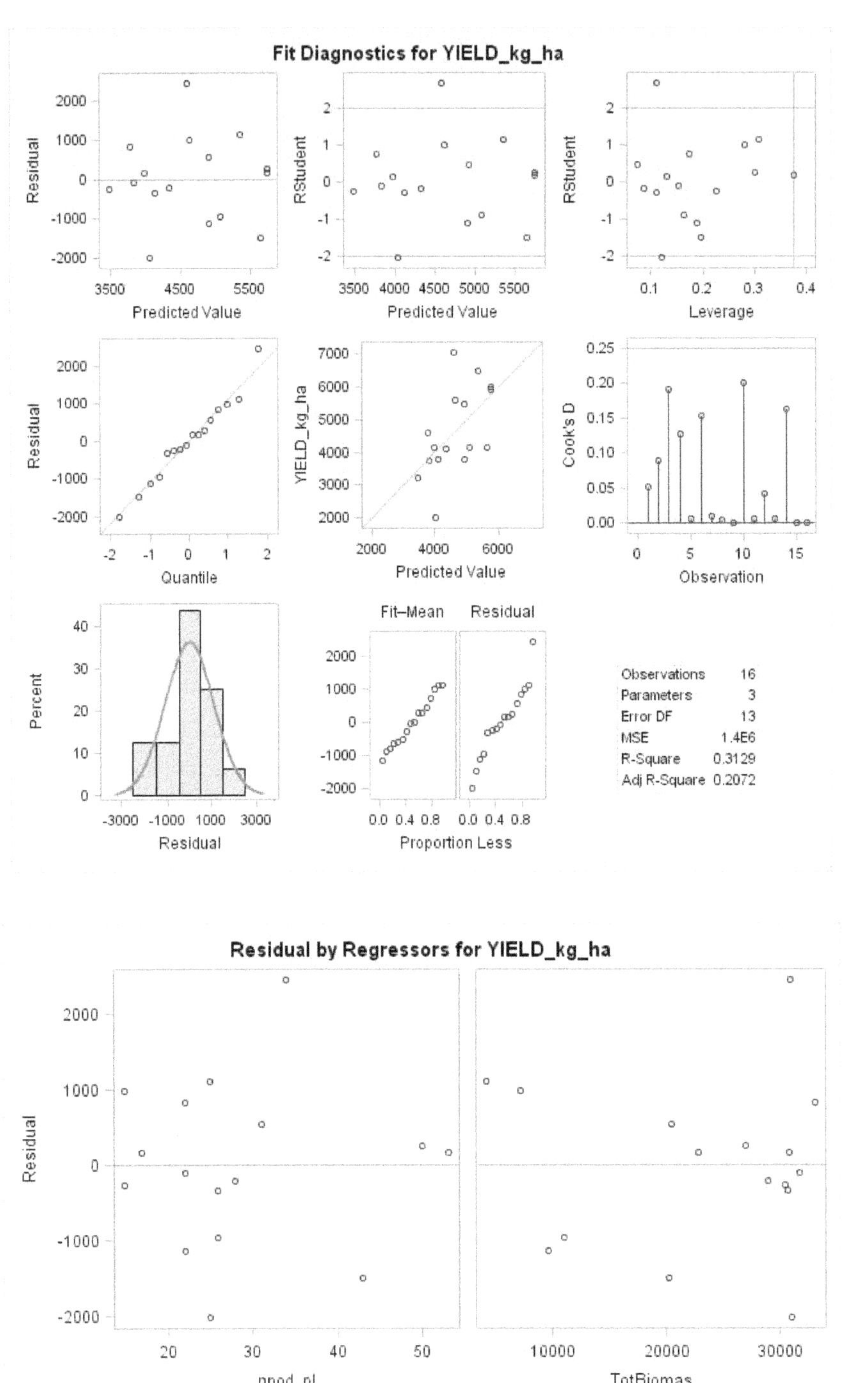

The SAS System

The REG Procedure

Model: eq1

Dependent Variable: YIELD_kg_ha

plant=PigeonPe

Number of Observations Read 16

Number of Observations Used 16

Analysis of Variance

Source	DF	Sum of Squares	Mean Square	F Value	Pr > F
Model	1	2702710	2702710	1.28	0.2777
Error	14	29657307	2118379		
Corrected Total	15	32360017			

Root MSE	1455.46524	**R-Square**	0.0835
Dependent Mean	1819.28750	**Adj R-Sq**	0.0181
Coeff Var	80.00194		

Parameter Estimates

Variable	DF	Parameter Estimate	Standard Error	t Value	Pr > |t|
Intercept	1	1171.91317	678.88393	1.73	0.1063

| N pod pl | 1 | 30.73587 | 27.21120 | 1.13 | 0.2777 |

The REG Procedure

Model: eq1

Dependent Variable: YIELD_kg_ha

plant=PigeonPe

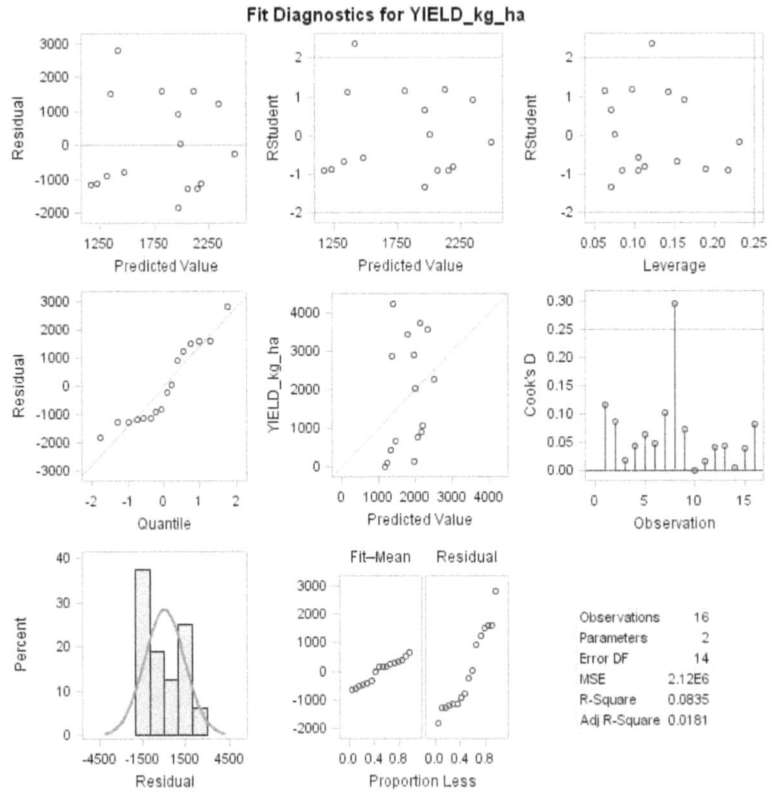

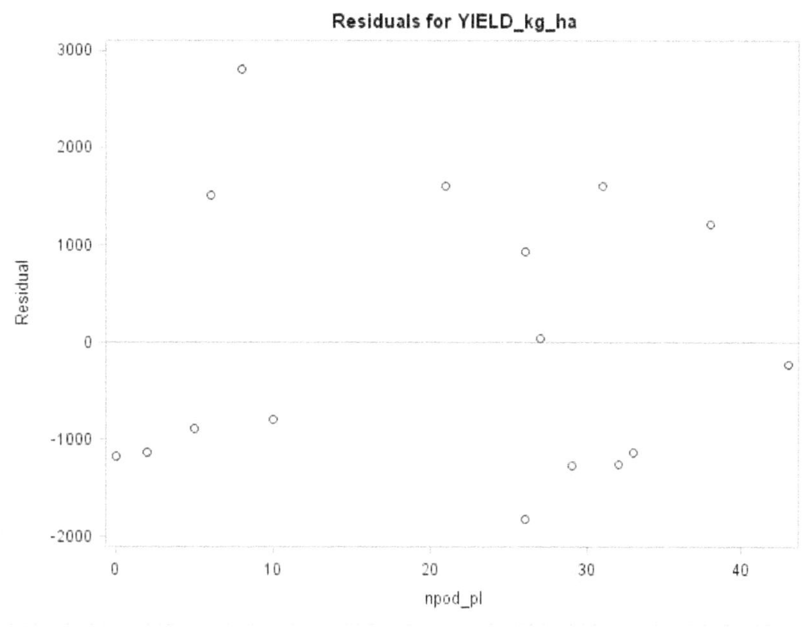

Residuals for YIELD_kg_ha
Residual
3000
2000
1000
0
-1000
-2000
0
10
20
30
40
npod_pl

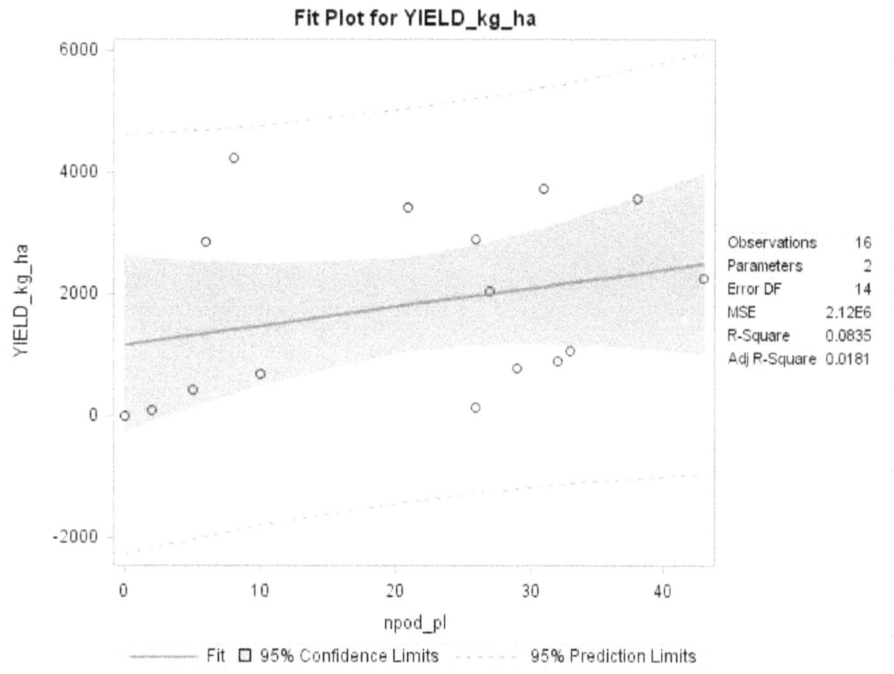

Fit Plot for YIELD_kg_ha
YIELD_kg_ha
6000
4000
2000
0
-2000
0
10
20
30
40
npod_pl
Observations 16
Parameters 2
Error DF 14
MSE 2.12E6
R-Square 0.0835
Adj R-Square 0.0181
Fit
95% Confidence Limits
95% Prediction Limits